Mysterious artifacts

Tadeusz Oszubski

Scientists analyze the history of mankind, determine their course and meanders. They supplement our knowledge of the past by discovering and describing further artifacts, that is, objects that are the work of human hands. Texts written on parchment and carved in stone, works of art, everyday objects, machines, buildings – all monuments make up the image of civilizational development of our species. However, there are finds that do not match the findings of the researchers. Objects that either hide secrets or are puzzles themselves that science cannot solve.

These mysterious artifacts are a challenge and for the academic order, and for our imagination.

Voynich Manuscript

The history of the book, known as the "Voynich manuscript", is mixed with a rowdy legend. Who else, then, as not a man like Michal Wojnicz, could reveal to the world the existence of this mysterious work?

Michał Wojnicz of Habdank coat of arms was a Polish nobleman and a child of his era. Born on October 31, 1865 in Telsa, Lithuania, he grew up in a patriotic and highly political environment.

Poland was enslaved at the time. Militarily and economically weakened Poland was invaded and occupied by three border states, Prussia (today's Germany), Austria and Russia. This process lasted from 1772 to 1795 and is called the three partitions of Poland, which regained statehood only after 125 years, when World War I ended.

Wojnicz lived in an area of Poland annexed by Russia, which fought against all manifestations of patriotism and attempts to regain freedom by Poles. In 1863, the Polish so-called January Uprising broke out in the lands under Russian rule, bloodily suppressed by the Tsarist authorities. Michał Wojnicz, growing up among the memories of patriots, heroes of the uprising, joined as a young man in the ranks of the secret revolutionary organization

Proletaryat founded by Ludwik Waryński. Hovewer, before Wojnicz's career as a conspirator and dynamitard could develop, in 1885 he was arrested by Ochrana, the tsarist political police. At that time, Russia did not caress its enemies. The process didn't take long. A Pole was sentenced to exile for his political activities. Like many young people of his generation, he ended up in Siberia. Except that Habdank did not intend to submit to the will of power. It is difficult to say today what weighed on his decision. Whether it was patriotic motives, hostility towards the possessor, aversion to difficult living conditions, or, above all, a desire for adventure. In 1890, 25-year-old Michael escaped from the exile and then crossed Asia from the freezing north to its tropical south.

After an adventurous journey, he made his way to London. There he adapted the record of a difficult Polish surname to English pronunciation. His name was Wilfrid Michael Voynich. It may have been one of his treatments to hide from Ochrana's agents. In London, he took up a new profession and found love. He ran an antique shop, and his wife was English woman Ethel Lilian from the Boole house, with whom he spent the rest of his life. It can be said that from now on the former conspirator, then a brawler and a traveler, led a banal life in the family circle, trading dusty books. All this, however, changed in 1912, when during one of Voynich's commercial trips to Europe he found himself eye to eye with a mysterious manuscript. He bought it along with other old books from the Jesuits from Villa Mondragone in Frascati, Italy.

At first glance, the mysterious book resembles the almanacs written by alchemists in the 16th and 17th centuries: it is framed in leather, measures 15 by 22.5 centimeters, and is written on veneer parchment. To date, 120 cards from the original 136 have survived. The manuscript is divided into parts, and these into chapters. From the illustrations with which the book is richly bearing, it can be concluded that it is devoted to astronomy or astrology, besides herbalism, hydrofoil and other fields of natural medicine. However, the similarities to any known astronomical or medical work end.

First, the text was fixed on parchment by an unknown letter. Secondly, it is not known what the illustrations represent, the interpretation of which has become a challenge – first for Voynich himself, and then for the multitude of scholars and lovers of esoteric knowledge. To this day, it is not easy to reveal the mysteries of the book, which is not even known who wrote it and what title it bears. For these reasons, the manuscript is referred to as the "Voynich manuscript".

In 1914, Wilfrid Michael Voynich moved with his wife to the United States, where he began to run an antique shop in New York. He died on March 19, 1930. Until recent days, he tried to decipher the contents of the most enigmatic book he had to deal with.

In 1960, the antiquarian's widow donated the manuscript to the library of Yale University, where it is still stored today, spending a dream with generations of mystery enthusiasts, bibliophiles and scientists – historians and cryptologists.

They all ask the same questions: who wrote Voynich's manuscript, when and for what purpose?

When was this work created? The first mention of it dates back to 1585, when John Dee sold it to Emperor Rudolf II (ruler of Germany from 1575 to 1612) for 600 ducats. Rudolf Reided in Prague, Czech Republic, and was interested in arcane knowledge. He supported the work of alchemists and astrologers, and collected unusual objects such as this unreadable manuscript. After Rudolf's death, the book was placed in the Jesuit library in Frascati under unclear circumstances, from where Voynich bought it.

What does the manuscript contain? According to some mystery researchers, it is a book of Templars, containing descriptions of their esoteric practices and the location of hidden treasures. Another concept is that it is the sacred text of the Qataris that survived the Catholic Church's crackdown on these medieval heretics. Another hypothesis is that Voynich's manuscript is an encrypted compendium of knowledge written by one of the great thinkers of the old centuries, such as Roger Bacon or Leonardo da Vinci.

However, the two exchanged various handwritten notes, the analysis of which indicates that the manuscript is not their work.

It is true that da Vinci betrayed a weakness for ciphers and encoded some of his texts in mirror writing, but none of the documents he produced showed signs that were written Voynich's manuscript. So are these characters an extremely hard-to-crack cipher, or are they rather "letters" of a forgotten alphabet used to write some extinct language? Physicist at Yale University, Dr. William Ralph Bennett, stated in 1976 that the text was related to the languages of Polynesia, and according to Brazilian researcher J. Stolfle, it could be a phonetic record of the message in Chinese.

The history of the book was marked by another Polish accent. Zbigniew Banasik, an amateur linguist from Wroclaw, announced in the autumn of 2003 that he had broken the code of Voynich's manuscript. According to Banasik, the manuscript is a kind of vademecum – a collection of knowledge gathered by a more unspecified shaman, written earlier than evidenced by radiocarbon analyses of the monument, because between the 13th century BC and the 3rd century BC. Banasik bet on the 4th century BC.

"Comparing the different alphabets and languages, I found that there was a great similarity between the word used in the book and the Manchu language," Banasik said in a statement to the media. "I was able to read 45 characters of the alphabet, one of which I did not find in the text, and the compatibility of more than 50 words with The Manchu language.

This, however, ended the Polish contribution to unraveling the puzzle, because Banasik's bumous announcements did not go into concrete. The manuscript language researcher fell silent and never revealed whether he had read the entire text.

Another who claims to have broken the Voynich Manuscript code is Canadian researcher Amet Ardic. As October 18, 2018, he handed over the service "Ancient Origins", according to Ardic, who is of Turk origin, the book was written with signs reminiscent of the old Turkish script, but by phonetic spelling. Even the design of the words resembles the Turkish language used 500 years ago. The researcher claims to have discovered the meaning of about 300 words in manuscript and plans long-term research to read the full text of the book based on his assumptions.

There is also no shortage of critical opinions about Voynich's manuscript. In April 2007, Dr. Andreas Schinner, an Austrian theoretical physicist and programming expert at the University of Johannes Kepler, presented the results of a computer analysis of the text of the book. According to Schinner, statistical analyses show that the characters with which the book is written do not convey any content.

"It's gibberish," said the Austrian. – The statistical characteristics of the text indicate that we are dealing with a joke or a scam.

Earlier, in 2004, Dr Gordon Rugg of Keele University in the UK came to a similar conclusion. According to him, Voynich's manuscript is the work of a fraudster.

"Just because there are subtle regularities in the text of the structure and distribution of words, no one has ever considered the possibility that this work is an ordinary hoax," the Briton said in an interview.

It is worth mentioning that Rugg's last proposal is not true, as it was already in the 1970s. In the 20th century, Dr. Robert S. Brumbaugh of Yale University decided that the manuscript was not authentic. According to this researcher, the book was made only to extort as much as possible from Emperor Rudolf, who was widely known to be a collector of mysterious artifacts. If the manuscript is actually the work of a madman, who is responsible for its creation?

The first suspect, of course, is the man who sold the manuscript to the German ruler, John Dee. This is a special character, because even the rowdy fate of Michał Wojnicz-Voynich falls pale in confrontation with the history of John Dee.

The life of this Englishman born in 1527 and who died in 1608 or 1609 is the material for a fantastic-adventure novel. In his biography, historical facts mingle with a legend that over time grew around the figure of an astrologer, alchemist, magician and spy of the English Queen Elizabeth I, reigning from 1558 to 1603. Apparently, in his studio near London, Dee evoked the angels and ghosts of dead people. In his actions he was to use the help of a crystal mirror and a crystal ball. John Dee, and therefore indirectly and with Voynich's manuscript, is associated with another Polish accent. Before the astrologer went to Prague to Rudolf's court to sell him the manuscript, he ended up in our country.

Together with the Polish magnate Olbracht Łaski and the English alchemist and medium Edward Kelley, he came to the court of Stefan Batory (reigned from 1576 to 1586), where he gave shows of magical skills and collected information for his queen. The

Polish stage of Dee and Kelley's sorcery-espionage adventure lasted several months, after which they both went to Prague.

According to sources from the era, John Dee was able to talk to angels. He and Kelley were also to write the "language of angels," otherwise known as Enochian scripture (after Enoch, one of the prophets of Israel who reportedly spoke face-to-face with Jehovah). Here you have to go back to the aforementioned analyses of Dr. Gordon Rugg.

"The most unusual feature of the Voynich manuscript is the mirror image of the most distinctive features of the Enochian language," said the British scientist. – The text of the manuscript accurately reflects what one would expect from an Enochian language creator if he wanted to verify the shortcomings of his original creation and create an even better illusion of natural language.

How did the Enochian language work? From 1582 to 1589, Dee worked with Kelley, who was considered a medium for contacting angelic beings. Dee recorded everything he heard from his accomplice's mouth while he was in a trance during the screenings.

When the alchemies systematized the notes, they received plaques with signs of Enochian script, still stored in the British Museum in London. Apparently, during one of the screenings, an angelic being named Galvah appeared, who announced that Dee would create a special work: a book written in a new world language to replace the Bible and give humanity a new law coming from heaven. Was it a Voynich manuscript?

Researchers say that Enochian words do not resemble words from any human language. So what is this supposed language of angels? Just a gibberish used to deceive gullible sponsors, or perhaps proof of the existence of metaphysical beings living in another dimension?

One can also ask whether the language of intangible beings must be governed by the same laws as human speech.

On this, however, angelological considerations must be concluded, because the existence of angels is based on faith, and there is therefore no empirical evidence of their existence.

Historians determined that Dee sold Voynich's manuscript to Emperor Rudolf for 600 ducats. A lot or not enough? Books in that era cost a lot, for example, in England, for the important legal work A Grand Abridgment of the Common and Statute Law of England had to be paid – in terms of the currency used in continental Europe at the time – roughly 2 ducats. The new ambulance cost around 40 ducats in London and a good quality gentleman's outfit cost 20 ducats. If Voynich's manuscript is a hoax, it could take John Dee just three months to write and decorate it with illustrations, meaning it could be said that the Queen's spy did the business of life. On the other hand, it is necessary to remember the findings of Gordon Rugg, associating the manuscript with the Enochian language, whose creators were Dee and Kelley. So maybe the book is not so much a forgery as a mystical work actually written in the revealed speech of angels and depicting another world?

The Enochian writing should have been mentioned here, because it was not forgotten even four centuries after Dee's death.

The alchemist's journal records were based on some of his research by the creators of the esoteric currents of the turn of the 19th and 20th centuries – Wynn Westcott and S. L. MacGregor Mathers, founders of the Hermetic Order of golden dawn. He also studied the language of angels in the 1920s. In the 20th century, another well-known esoteric, Aleister Crowley.

To this day, Voynich's manuscript is handled by both scientists and enthusiasts approaching the manuscript by paranormal methods. The latter group belongs to a certain Veikko Latvala, a businessman and visionary from Finland, who in March 2012 assured the world that he had unraveled the mystery of the manuscript.

Latvala describes himself as God's prophet and, he claims, it was through divine help that he managed to decode one of the world's most mysterious books. The words of the Finnish prophet are written down and communicated to the world by Ari Kepol, previously his partner in business ventures.

"The characters in Voynich's manuscript describe sound waves and vocal syllables," explained the mystery of Ari Kepol's manuscript. – The language used in this book is complex. The sound syllables are written in a mixture of Spanish and Italian. Other passages are equally patchy. The complication of writing is due to the language used by the author of the manuscript. He spoke in the Babylonian dialect used in a small area of the Middle East. He didn't know how to perpetuate his knowledge, so he had to create his own alphabet and vocabulary. At the same time, the record takes the form of abbreviations, like those used in shorthand.

But what does the enigmatic book say, according to Latvala?

"It is the work of a life forgotten today," said Ketola, passing on the findings of the Finnish visionary. "This is a scientific medical publication, and the knowledge contained in it will be useful today. The writer was a scientist: botanist, apothecary, astrologer and astronomer. They know he's been ahead of his time by decades, or even hundreds of years. This book also contains prophecies.

In other words, Voynich's manuscript is primarily intended to be a work concerning the use of plants in medicine and other branches of science. According to Latvala, the book describes 16,152 plants. An example is to be growing to this day in Ethiopia – as the visionary assured – the heart of fire, which "makes the skin beautiful", from the oil squeezed from the buds of the Heart of Fire is made ointment that eliminates wrinkles, and the infusion of its flowers prevents inflammation, that is, it is an antibiotic. This plant is to be dry and heat-like and 10 centimeters high and light green. So far, scientists have not discovered such an herb, but according to Latvali, there will come a time when this will happen.

For almost a hundred years, a legion of cryptologists tried to decipher the manuscript. To no avail. So how did Finn do it? The answer is simple: Latvala has direct contact with God.

"An ordinary person would not be able to decode this book because there is no code or method to read the text contained in it," he explained for his accomplice Ketola. –This is due to the fact that the manuscript is written in the language of the metaphysical channel of prophecies. People with the gift of receiving this channel are rarely born, but for millennia one of them has always lived on Earth. Now this role is played by Mr. Veikko Latvala, who obtained this gift 20 years ago.

Cryptologists approach Fina's findings with great caution and refuse to comment on his conclusions, as he has not used any scientific method and draws his knowledge from metaphysical visions. Moreover, Latvala is not the first to use this approach to uncover the mystery of the manuscript.

As you can see, scientists and amateurs of mysteries struggled with Voynich's manuscript.

Some of them hypothesized about the form of the record and the content, but no one presented a reliable translation of the book. This has led to the belief in academia that the manuscript is a forgery and not worth bothering with. Nevertheless, several scholars over the past decades have conducted computer analyses of characters from the book cards. Most of them stated that the manuscript does not have the correctness typical of existing languages or forms of encryption. Thus, they confirmed what scientific circles have taken for granted before: we are dealing with a hoax.

What's different about the credibility of Voynich's manuscript is believed by two experts who analyzed this artifact using modern technology. In June 2013, the scientific service "Live Science" reported on the results of the work of Doctors Marcelo Montemurro and Damian Zanette, physicist at the University of Manchester. They found that the letter used on the pages of the book was not forgery, because the characters making it up were statistically organized in a way that was characteristic of the real languages.

"Although the mysteries of the origin and meaning of the text are still unresolved, the evidence gathered on the structure of the organization of that text, at various levels, indicates that it has a genuine linguistic structure. This undermines the opinion that Voynich's manuscript was falsified," Montemurro said.

These findings were supported by other researchers – in January 2014, Dr. Arthur O. Tucker, a botanist at Delaware State University, and Rexford H. Talbert, an engineer working for NASA and the U.S. Department of Defense, presented a new hypothesis about the manuscript. According to them, it was created in Central America.

Tucker, an expert on ancient Mexican herbaceionage, determined what kind of plants were depicted on the pages of the book. Until now, it was assumed that the flowers, herbs and shrubs painted there were fetuses of the imagination of an unknown author, but Tucker proved that some of them look the same as the plants depicted in the illustrations of a Spanish manuscript drawn up in the 16th century in Mexico. Tucker and Talbert found that thirty-seven plants, six animals, and one mineral, as shown in the manuscript, have their exact counterparts in the illustrations written in 1552 in Latin for the Code de la Cruz and Badianus. This code is a medical book containing information on herbs and natural treatments used by the Aztecs and other Central American Indians. Both researchers hypothesized that Voynich's manuscript was written in signs of a forgotten writing today, which once perpetuated the now extinct dialect of Nahuatl (whose other dialects are still used in Mexico today). Importantly, in this case, we are dealing not with another speculation of enthusiasts of the mysteries of the old world, but with a concept with a solid scientific basis, which was supported by many researchers.

"Previous attempts to crack Voynich manuscript code have focused on linguistics and cryptography. Tucker and Talbert focused on botany and convincingly identified many plants that are found on the American continent, commented Dr. Wendy Applequist, a specialist in American flora at the Missouri State Botanical Garden.

Other scholars, including Dr. Jules Janick, a botanist at Purdue University in Indiana, express similar opinions about Talbert and Tucker's discovery.

"Their analyses are groundbreaking in that they accurately show analogies to Mexican flora and fauna, which in the 16th century were handled by Spanish priests based on information from the Aztec elite," Janick said. – The names of the plants indicated have already been identified in Nahuatl. It is true that the vast majority of the manuscript is still undiciphered, but since there has been a breakthrough, I am optimistic about the near decoding of this book.

The Tucker-Talbert hypothesis indicates the direction of research to linguists and cryptologists. It seems that the focus of experts on Central American culture will make it possible to reveal the mystery of Voynich's manuscript. However, there was a problem: namely, the dating of parchment proves that the manuscript was written long before the discovery of America.

The first attempts to date the book using radiocarbon C14, carried out in the 1970s and 1990s, were carried out in 1998. Of the 20th century, they showed that the parchment on which the manuscript was written was made between 1300 and 1660. These findings, however, lacked precision, but at least undermined the long-standing opinion that the manuscript was a forgery of the early 20th century, perhaps even the work of Voynich himself. Subsequent analyses conducted in 2009 by experts at Arizona State University in Tucson found that the components for the preparation of the book (parchment, ink, paint) were made between 1404 and 1438. The manuscript was re-researched at the same university in February 2011. State-of-the-art equipment from the NSF Arizona Accelerator Mass Spectrometry specialist resource was used. Four parchment samples from four randomly selected pages of the book were analyzed, which ensured that the results obtained would relate to the entire artifact.

Using a radiocarbon dating method, a team led by Dr. Greg Hodgins, a chemist and archaeologist, further determined the time it was to produce parchment.

All samples were found to be from between 1410 and 1430. And that's the problem, because the areas of the Aztec state in Central America were occupied by the Spaniards only a hundred years later.

We already know, therefore, that the dating of the material on which the book was written is not in doubt from the scientific point of view. The hypothesis about the Mexican origin of the artifact also has strong points. And since parchment was an expensive thing, there is no way that after its manufacture the cards lie unsaved for a whole century. Someone used this material – and probably right away. It follows that Voynich's manuscript was written one hundred years before Herman Cortez conquered the Aztec state. Are Tucker and Talbert right? If so, maybe someone has past Columbus in discovering America, and the book is the yield of an unknown historian's journey to the New World.

Did the mysterious explorer sailor have the medical knowledge of the Aztecs in the manuscript? This is speculation, of course, but you can get the impression that everything is possible with this mysterious artifact. However, the book consistently protects its secrets. As soon as a fact is established about it, another unknown immediately appears.

Machine from Andikithira

The device called the Andikithira machine is the size of a shoe box. Although small and complex, the mechanisms with a similar degree of complexity were not built in Europe until the 18th and 19th centuries, and these were multifunctional clocks. However, this artifact was built in antiquity. It can be said that it was two thousand years too early.

The machine from Andikithira was discovered in 1900 in the wreck of an ancient merchant ship whose remains rested at a depth of 60 meters in the sea off the coast of the Greek islet of Andikithira (also known in the sources as Antikitera or Antikythera). Underwater archaeology at that time did not yet exist, and equipment was not known to work efficiently at high depths. Greek divers, relying solely on the efficiency of their lungs, struggled to reach the bottom. There, they grabbed what hit their hands and hurriedly returned to the surface. Nevertheless, an archaeological team led by Valerios Stais has amassed a number of artifacts that once filled the holds of an ancient ship. On the basis of analyses of the finds, Stais concluded that the craft sank in the 1st century BC.

The finds were taken to the National Museum of Archaeology in Athens and in their establishments.

Only one artifact that did not match the scientific knowledge of ancient times at the time proved troublesome. Experts at the beginning of the 20th century originally recognized that it was a kind of mechanical calendar, after which they stopped bothering with it.

This is how the machine from Andikithira ended up in the museum magazine. And there remained long decades. For two-thirds of a century, science celebrated this find from afar. Dr. Derek de Sola-Price, a lecturer at Yale University, became interested in the first. In the 1960s and 1990s, the in the 20th century, he examined the device and, putting his scientific authority at stake, ruled that it was an incomplete, shipwrecked counting machine. In other words, the mechanical computer of the ancient era.

The researcher assumed, workingly, that the machine was used to calculate the movements of the Sun and moon, which allowed to determine the geographical position. So it could serve both sailors and people traveling on land. De Soli-Price's conclusions were criticized by the academic community at the time.

Skeptics argued that no ancient source known to science gave or described mathematical principles for building such complex devices. The machine from Andikithira was again covered with the dust of oblivion.

The breakthrough came only in the new millennium, when a discovery was made that provoked scientists to devote energy and resources to reconstructing an alleged ancient computer. An avalanche of events was triggered by the examination of a prayer book from the 13th century, the text of which turned out to be a palimpsest, a manuscript written on parchment, from which the original text was taken or stained earlier in order to reuse expensive parchment material. At the turn of the 20th and 21st centuries, there were already modern techniques for reading the older record without destroying the document itself (earlier methods of reading palimpsests, used since the 19th century, usually irreparably damaged manuscripts). In December 2003, Dr. Reviel Netz, a mathematics historian at Stanford University, reported that the original text of the book, hidden beneath 13th-century prayers, was a 10th-century copy of Stomachion, Archimedes's scientific work that was thought to be missing.

Archimedes lived in the 3rd century BC, in the Greek city of Syracuse in Sicily. He made many discoveries in the field of geometry, arithmetic, mechanics and hydrostatics. He formulated the so-called Archimedes law, the principle of lever action, which, according to the anecdote, was to express in simple terms: "Give me a point of support, and I will move the Earth." He built a screw for pumping water, which is still used today to irrigate fields. He is considered the inventor of the multi-circle and water clock. He is credited with developing new combat machines, including catapult. However, he was remembered mainly for his achievements in mathematics. Many of the works of this ancient scientist have not been preserved to our time. It has long been believed that Stomachion, the combinatorial treaty, the field of mathematics, which today forms the basis of counting machines and computers, has also been lost forever.

Interesting were the history of the accidentally discovered copy of Stomachion. The manuscript was created around 975 and was one of three copies of this work by Archimedes known in the Middle Ages.

In the 13th century, the then owner (today we do not know who he was) apparently considered the content of the ancient treaty on mathematics to be irrelevant, so he transferred it for religious purposes. The book was cut into new pages, then the record was taken from parchment, and the purified material was used to make a prayer book.

For more than half a millennium he passed from hand to hand, and finally went to one of the libraries in Istanbul.

In 1906, he was stumbled upon by the Danish scholar Johan Ludvig Heiberg, who saw traces of an older text with mathematical symbols under verses of prayer. Using a magnifying glass, he rewrited two-thirds of the manuscript, and came up with the idea of photographing the original pages. Unfortunately, he failed to decipher mathematical content. The following years brought several wars in that region, so the research work did not continue. Finally, in the 1920s and 1990s, in the 20th century, a Frenchman acquired the palimpsest with the Archimedes Treaty. Much later, in 1998, his heirs sold the book for two million dollars to an anonymous buyer who donated it to the Walters Art Museum in Baltimore, where it is still located today.

It turned out that the prayer book contained only part of the original Stomachion, moreover, historians were not sure what the preserved passage was about. Initially, it was assumed that this was a description of a children's puzzle called stomachion, which consists in laying triangles from papyrus or from pieces of bone so that different shapes are formed. However, the aforementioned Doctor Reviel Netz found that Archimedes tried in Stomachion to answer the question of how fourteen different triangles could be laid to obtain a square. Netza was supported by a team of four combinatorial specialists. For a month and a half, they worked on super-high-performance computers – and found that Archimedes' problem had 17,152 solutions. However, before these four mathematicians sat down to analyze the text, a fragment of the work had to be extracted from the top layer of the palimpsest. Ultraviolet lighting was used for this purpose, but the reading of the original recording was hampered by mechanical defects, i.e. holes, sharpened edges of the pages, fragments fought by mold.

It was necessary to preserve and clean parchment, and then come up with methods of reconstruction of the text. The latter task was undertaken by a combined force of American specialists from the Institute of Technology in Rochester, Boeing and Johns Hopkins University. Text reading programs have been developed.

Knowing the preserved part of Stomachion therefore required many maintenance procedures, the involvement of philological knowledge and computer technology.

Let's go back to Reviel Netz's research. In 2003, he discovered that the manuscript refers to combinatorics, a field of mathematics that was previously thought to have been developed only in the 20th century, mainly due to the work of computer programmers.

Over the following years, thanks to the tools of the aforementioned American mathematicians and programmers, Dr. Netz's team read most of the preserved content of the treaty.

And that's when Stomachion's relationship with the Machine of Andikithira became apparent. Netz's analysis showed that Archimedes built the theoretical foundations of the combinatorics on the basis of an ancient puzzle. So they began to consider whether the device extracted from the wreckage could have been the result of the practical application of the ancient scientist's theory. What's more, Archimedes was an excellent mechanic, so maybe he personally constructed a puzzling mechanism? Archimedes was killed by the Romans in 212 BC, according to the latest findings, the device was built just after 205 BC (the issue of dating is yet to be discussed), while the merchant ship on board the machine sank between 85 and 60 BC.

"The device may have been very old at the time of the shipwreck," Dr. James Evans, a science historian at the University of Puget Sound in Washington, said in a statement to the New York Times in November 2014. "The new date of construction of this "antique computer" indicates that it is unlikely to be directly connected to Archimedes."

So who built the machine? If not Archimedes, then maybe one of his disciples or another of the Greek scientists of that era, for example Hipparch or Poseidonios? Let us remember that only small fragments of literature have been preserved from ancient times. The vast majority of documents, scientific and literary works have been lost irretrievably. The name of the machine builder from Andikithira and the description of his work may also have disappeared from the history books. However, even putting aside the riddle by this device, the second mystery does not give peace. What was the complicated mechanism used for? The answer is not facilitated by the fact that only one, and it is badly damaged, a copy of this device is known. Is it a unique, individual invention? Or maybe such calculation machines were popular in the days of antiquity, and only because of the lack of luck archaeologists have not yet come across other copies or even their descriptions?

First, when was the device properly built? Preliminary findings, made back in 1900 by Valerios Stais, indicated the 1st century BC. However, a closer date was not indicated until November 2014. This was the result of a research program led by the aforementioned Dr. James Evans, a historian of science. After only a few months of work, the results of preliminary analyses of the front part of the artifact were published, where the remains of two concentric circles with zodiac signs connected to the internal elements of the device were found.

"It's a very sophisticated part of the machinery," Evans said in the spring of 2011. "No one imagined that there might have been something so complex in the 2nd century BC.

This researcher analyzed selected fragments of the machine stored for more than a hundred years at the National Archaeological Museum in Athens. Only a few parts of the mechanism remained in good condition, others were damaged by corrosion so much that they could neither be identified nor reproduced. According to Evans, the device had very precise gears to reproduce the apparent movement of the moon, sun and planets from our system: Mercury, Venus, Mars, Jupiter and Saturn. Because the orbits of these planets and the Moon are elliptical, they also could not simulate circles in the machine of celestial bodies, because the markings from the artifact would not match what is visible in the night sky. For now, there is no physical evidence to support this scientist's hypothesis, but if he had proved the existence of "planetary gears" in the device, he would have also explained some of the conundrums of the machine's operation.

For the next three years, Evans continued to study the mechanism, which he did in collaboration with Dr. Christian C. Carman, a historian of science at Quilmes National University in Argentina. On November 26, 2014, the two scholars published their final findings on the smithsonian magazine website. According to them, the shield at the back of the mechanism reproduced the solar eclipse that occurred on May 12, 205 BC. Such a proposal pushed the date of construction of the machine to the period after May 205 BC (however, it remains an open question whether it happened just after an unusual astronomical phenomenon, or 20, 80 or 100 years later – on the basis of records of this anaesth induch). It has also been shown that in predicting solar eclipses, the constructor of the device relied on Babylonian arithmetic rather than Greek trigonometry, which is extremely interesting, because the machine was built in the Greek environment, not mesopotam.

The results of analyses carried out at the same time by Dr. Paul Iversen of Case Western Reserve University in Ohio should also be added here. The work of this scholar allowed to recreate the inscription on the device referring to sports competitions held in Rhodes. According to Iversen, this may indicate that the machine is from this island.

All these new findings and hypotheses were made possible by an earlier scientific project that was completed in November 2006. It was then that an international team of scholars reconstructed the machine from Andikithira, building a copy of it from brass and wood. Step-by-step experts discovered how the mechanism work. After all, it consisted of dozens of elements, including wheels and bronze modes, which exactly got entwined with each other. However, for the machine to work, each part had to go to the right place. So they tested the original machine fragments using computed tomography, which was done by British scientists at Cardiff University working under the direction of Drs Mike Edmunds and Tony Freeth. The analyses were carried out under the auspices of the universities of Athens and Thessaloniki and the National Archaeological Museum in Athens. The work showed that 82 original parts were preserved, including 37 bronze gears, and the mechanism was driven by a hand crank. The reconstruction was based on elements in the museum's collections, but it is not known how many parts are missing.

The reproduced machinery turned out to work, but we do not know if it was possible to preserve all the computing capabilities of the original device. Making a copy of the "antique computer", although some parts are believed to be missing, has accelerated the research. In 2008, another function of the device was discovered. A team of American and British experts, working as part of the Andikithira Machine Research Project and led by Dr Tony Freeth, proved that the device could also be used to calculate the dates of the four major old Greek panhellene sports games, which were held in Anti-Olympic times in Olympia, Delphi, Corinth and Nemei.

The Machine from Andikithira is one of the finds that seem to be anachronism, at least from a science point of view. It is still unclear who invented and built the complex mechanism ahead of similar devices by two millennia, or how many of its units were made. It is also not fully known what this artifact was used for. The machine puzzle is still waiting to be solved.

Roman 12-beans

Artefacts called Roman dodenitis hide so many puzzles that scientists have made many hypotheses about the purpose of these monuments. Archaeologists and historians are still arguing over whether these are fragments of some mechanical weapon, toys, weather change meters, astronomical instruments, or perhaps iconic objects serving in religious rites.

What has been possible over the past decades, since the discovery of the first doudenum, about these mysterious objects? Not much, and there is only one thing – that there were quite a few copies of these objects, because archaeologists found as many as a few dozen of them. Most in France, Switzerland and the western part of Germany, that is, in the areas of Gaul and Rhine Germania, lands inhabited in antiquity by the Celts and Germans. All artifacts of this type date back to the 2nd and 3rd centuries AD, when some of these worlds were part of the Roman Empire and some were heavily influenced by the Empire. Monuments have a regular spatial form with twelve walls, and they were usually made of bronze, less often made of stone. They are not large, because the length of their one side is from 4 to 11 centimeters. Some have knob-shaped tabs, but it is not known whether these elements have a decorative meaning or are due to a function unknown to us today.

The ancient Romans left behind a lot of works in which they meticulously described their daily lives and commonly used objects and devices. However, on the subject of dodenoths, Roman sources remain silent.

According to a new hypothesis on the destiny of the twelve-page, presented by Dr. Andrea Galdy, lecturer in art history and archaeology at the University of Manchester, the trace of the artifacts of interest to us here may be in one of the works of the Greek historian Plutarch (I-II centuries A.D.).

On June 10, 2011, news site Fox News reported that, according to Galdy Plutarch, he was mentioning doudenitis as the primary instrument for calculating the position of zodiac signs. Therefore, an English scientist assumed that these objects were instruments used in astrological calculations, and astrology was very popular in Gaul II and III centuries AD. The archaeologist also suggested that the symbols of the twelve zodiac signs were embedded in the empty spaces of the artefacts, which were not preserved because they were made of volatile materials such as papyrus, parchment or fabrics.

"I hope that this explanation, thanks to Plutarch's testimony, will help to organize our knowledge of these mysterious objects," said Galdy.

History researchers have taken this hypothesis with interest. As some of them point out, the assumption that 12th-century instruments may have been astrological instruments or fragments of larger devices can be taken seriously.

In Roman times, astrology, horoscopes and other treatments related to predicting the future were extremely popular. However, it is not known whether Galdy's hypothesis will be confirmed by other historical sources or new archaeological discoveries. And only in this case can it be unequivocally stated that Roman duodenals are objects used for astrological practices in the late antiquity era.

Starving from Sakkara

Some archaeological finds cause confusion in scholars, as it is not known how to interpret them. This is the object of a sculpture called "Starving of Saqqara".

Proponents of paleoastronautics argue that the Starving are another argument in favour of their thesis about contacts between extraterrestrial civilizations and our ancestors that have influenced the development of human civilization.

According to this concept, some finds should be interpreted differently than scientists do, for example, many ancient paintings and sculptures should be considered not as representations of deities or famous ancestors, but aliens who came from outer space.

The most famous popularizers of paleoastronautic theory are Erich von Däniken and the recently deceased Zacharia Sitchin.

However, before we succumb to the suggestions of supporters of contacts with strangers, we should determine what is officially known about the sculpture of the Starving.

It is sculpted in limestone, weighs about 80 kilograms and is 67 centimeters tall. It depicts two faces facing each other naked figures of unidentified sex, very lean and with heads of unusual shape. The style in which the sculpture was made is not known from the canons of art of any of the ancient cultures of the Old World. There is an inscription on the back of one of the figures – but it also does not explain anything, because it was fixed with signs of a mysterious handwriting, which scientists have not come into contact with anywhere else.

Can some light on the artifact shed light on the location and circumstances of the sculpture's discovery? Maybe that would allow it to be interpreted? Unfortunately, it is only known that it was discovered in the course of archaeological works probably carried out in The Egyptian Sakkara, which are not yet in the scientific report. Presumably, these excavations took place before World War II. Such vague information does not allow to draw almost any conclusions – it follows only from them that the sculpture comes from the territory of Egypt. Its execution time and purpose remain undetinated.

It came to Canada in 1940, and was brought there by Vincent and Olga Diniacopoulos, Greek immigrants from France who had previously collected a world-class collection of two thousand ancient monuments from Egypt and Israel, among others.

Many of these items, including the Starving of Sakkara, were exhibited in the 1950s at the Diniacopoulos Family Gallery of Classical Art, a large private museum in Montreal. In 1999, the late couple's son, Dr. Denis Diniacopoulos, a concordia University researcher, donated the family's collections to many cultural institutions around the world, including the Royal Ontario Museum in Toronto. Several exhibits, including Starving from Sakkara, went to Concordia University in Montreal. The sculpture was kept in a warehouse for more than a decade, but then the university authorities decided to put it on public view. Since mid-March 2011, the Hungry have been exposed in the atrium of the Concordia Institute of Engineering, Informatics and Visual Arts building.

"This facility is a unique work of art," Dr. Clarence Epstein, director of the university's Center for Culture and Special Projects, said in a statement to Canadian media. "We have put it on public view in the hope that it will attract the attention of an international group of researchers, which will help explain its origin. Over the past decade, our scientists have tried to determine the age of the sculpture and determine in what artistic tradition it was made. Unsuccessfully. That's why we contacted the British Museum, the Brooklyn Museum, the National Museum of Israel and the universities of Cambridge and Oxford. Even in these great world centers of archaeology, no one was able to explain the mystery of this sculpture. No one could translate the inscription ched on it. It was not made either in Aramaic, Hebrew, or in Egyptian demotic writing. Our monument remains a mystery to experts.

Some scholars assume that the sculpture is older than the civilization of ancient Egypt, and even that it was created thousands of years before the pyramids and is a trace of some lost culture of that region. Other experts accept that the Starving of Sakkara brought merchants from another region of the world to ancient Egypt – but from where, they no longer specify. All this, however, is only a presumption, because it is not known whether the statue was found in a preynastic tomb, or in a later burial, or in the ruins of one of the houses of ancient Sakkara.

So it is possible that we are dealing with both an "import" from another part of the world, as well as the only artifact that has survived after the still unknown science of civilization that existed in North Africa or the Middle East in distant antiquity.

As already mentioned, paleoastronautics supporters also got interested in sculpture. Enthusiasts of ancient contacts with aliens say that the Hungry depict aliens or interspecies hybrids, the offspring of newcomers and humans. This is indicated by the unusual proportions of the bodies of figures carved in limestone, and above all the shape of their faces and skulls. In fact, the Hungry People of Saqqara do look like aliens - at least according to common ideas about aliens. Who's right? Archaeologists or supporters of paleoastronautics? It is certain that the Saqqara sculpture is one of the most interesting unresolved mysteries of the past - as Dr. Epstein said honestly, pointing to the helplessness of experts from world centers of archeology in the face of the secrets hidden by the statue of the Hungry.

People have traveled the world for many millennia, and ideas and inventions have traveled with them. In the 20th century, scientists developed systems for the chronology of cultural contacts between peoples and civilizations from different continents.

However, there are monuments that cast doubt on some of these findings. Paintings, sculptures, giant symbols carved into the ground. Puzzling artifacts may indicate that the inhabitants of ancient Europe, Asia and Australia reached America long before Columbus. In turn, other sailors and builders created a civilization on the subcontinent, which according to geologists was engulfed by the waters of the South China Sea twelve thousand years ago. In terms of contacts between ancient cultures and the existence of so-called lost civilizations, mystery chases the mystery, and the riddle gives rise to another mystery.

Not only Nazca

The most famous geoglyphs, or large paintings etched in the ground, are located on the desert plateau of Nazca in Peru. More than 13,000 lines create 800 forms there – geometric signs and animal silhouettes. Since 1927, when pilot Toribio Mejia Xespe flew over Nazca and told the world about the huge signs visible in the desert, a legion of scientists and mystery enthusiasts have been trying to uncover their secret.

New geoglyphs are still being found, even in the area most closely studied for these symbols, namely Nazca. At the end of April 2006, another giant drawing was found on a Peruvian plateau. The previously unknown geoglyph is 65 meters long and depicts a horned animal of an undetermined species. The discovery was made by a team of Japanese scientists, analyzing satellite images of the area.

An expedition headed by Dr. Masato Sakai, a lecturer at Yamagata University, set off for Nazca. Scientists have confirmed that what they saw earlier in satellite images is indeed a hitherto unknown mark engraved in the ground. It resembled ornamental motifs on clay monuments from this region, made between 100 BCE and the year 600 CE, i.e. from the period of the development of the local Nazca culture, the work of which is many other geoglyphs known to researchers.

The Japanese also explored the plateau in the following years with further success. As reported on July 8, 2015, the site "Ancient Origins", scientists from Yamagata discovered as many as 25 new geoglyphs in the Peruvian desert. These signs are located just 1.5 kilometers from the city of Nazca and are presumably older than the symbols previously studied. They were not noticed before the arrival of the Japanese expedition because they are difficult to see due to the destruction of the earth's surface by erosion. Researchers in Asia managed to identify them using a 3D scanning device.

The signs show animals - presumably llamas - and unidentified geometric images. They measure from 5 to 20 meters and are older than the previously studied Nazca geoglyphs, because they were made between 400 and 200 BC.

Dr. Masato Sakai's research team continued to work on Nazca for years to come. On November 18, 2019, the Daily Grail reported that researchers at Yamagata University had discovered another 142 geoglyphs. Japanese scientists using drones, laser measurement techniques, 3D mapping of the site, and artificial intelligence technology developed in IBM laboratories systematically studied a gigantic archaeological site, which is essentially Nazca.

These subsequent identified geoglyphs were made between 100 BC and 300 BC. The smallest is 5 meters long, and the largest of them 100 meters. They depict various symbols, human silhouettes and animals: birds, monkeys, fish, snakes, foxes, cats and camels.

The famous Nazca plateau is not the only area of South America where archaeologists have found geoglyphs.

In early March 2005, another giant potato rite was discovered near the town of Palpa – 50 kilometers from the Nazca Desert and about 350 kilometers south of Lima. These images were also found during the analysis of satellite imagery.

A team of scientists from the international Andean Institute for Archaeological Exploration, who have been working in the area for eight years, studied the newly spotted drawings. They, too, turned out to be older than the rites of Nazca.

Soon, in the area of Palpa (over 200 square kilometers), about a thousand geoglyphs depicting various patterns and human and animal figures were found. Since the images are only readable by an observer high above the ground, archaeologists from the Andean Institute working nearby had no idea of them. It was assumed that the authors of the rites were the creators of the Paracas culture, i.e. the Indians who lived in this region from the 7th to the 2nd century BCE. It is believed that it was from them that the creators of the Nazca culture took over the habit and skill of creating geoglyphs. It can therefore be assumed that in those territories there was an intercultural continuity of the ritual use of these signs over the course of almost 1.5 thousand years.

This is not the last area in Peru marked by geoglyphs. As reported on September 26, 2016, the scientific website "Live Science", previously unknown geoglyphs were also found in the Sihuas Valley, near the pre-Columbian trade route leading to the former city of Quilcapampa.

An international team of scientists led by Dr. Justin Jennings from the Royal Ontario Museum in Toronto found the find, which was surprising by its size and originality. According to the first findings, it is an irregular pattern carved into the ground made of rings of various sizes, ranging from 2 to 800 meters in diameter. It should be emphasized that this was the first time that such patterns were found in huge artifacts.

Geoglyphs in the Sihuas Valley were discovered in satellite images a year earlier. After collecting funds and a team of experts, the find was analyzed by conducting field research, also with the use of drones. Thanks to this work, it was established that the mysterious signs were engraved in the ground between 1050 and 1400 AD.

Another South American - but far away from Nazca and Palpa - geoglyphic region is the Bolivian plateau surrounding Mount Sajama, which is 6,542 meters high. In 2003, scientists discovered there an image complex that is now considered to be the largest group of terrestrial engravings in the world. According to a team of American archaeologists from the Pennsylvania State University, there are as many as 436 figures in the Sajama National Park, the largest of which is up to 20 kilometers. The lines that make up them are one to two meters wide. The Bolivian set of paintings covers an area of 22,000 square kilometers - an area sixteen times larger than the area occupied by the Nazca geoglyph complex.

American archaeologists went into the field a few months after the signs in the photographs were discovered. They assumed that the ground-based images of Sajama were created by a culture that flourished in that area between 200 BC. a 650 AD, which is about the same time as the Nazca culture. Perhaps, then, there were contacts between them - whether they were commercial, cultural, or warlike - and on their occasion there was a flow of ideas? Was it precisely in this way that the concept and technical skills of making geoglyphs were first transferred from the lands of the Paracas culture (in the Palpa area) to the Nazca cultural areas (Nazca plateau), and then from there to the Sajama plateau, where another Indian culture was developing?

From such an assumption it could appear that the inventors of the idea of geoglyphs were the creators of the Paracas culture.

However, the mystery of the origin of South American ground drawings will not be solved so quickly and easily, as much older carvings have been discovered in a different, and remote area of the continent.

Since 2012, scientists have found in the Amazon jungle more and more traces of a highly developed culture from the pre-Columbian era. Previously, it was assumed that the rainforest areas were inhabited only by tribes of hunters and gatherers who were unable to produce a higher culture. Among the new finds from the jungle, huge geoglyphs are particularly puzzling.

They have been encountered since the 1970s, but more attention was paid to these finds only in 1999, when enormous glyphs were noticed after cutting a large forest area (for the planned pastures for cattle). Similar finds have been made throughout the region. Scientists have rarely received reports of these random discoveries, and if anything, they were long delayed. More years passed before the first ground mark research was organized, found in an area of 250 kilometers on the border of Bolivia, Peru and the Brazilian state of Acre. As reported on February 16, 2015, the website "Ancient Origins", scientists from an international research project aimed at inventorying and examining Amazonian marks engraved in the ground, were given technologically advanced devices at their disposal.

Ancient symbols are overgrown with dense forest and are difficult to see, as well as reach them, so the work of scholars became more effective, it was decided to use drones. To this end, the European Science Council has allocated EUR 1.7 million to research carried out by experts in the United Kingdom. Drones equipped with laser measuring instruments, so-called lidars, were hired with funding from the European Union with the Brazilian authorities. Flying robots can locate and measure geoglyphs much faster and more precisely than people breaking through the jungle thicken, or even from teams moving planes or helicopters. Drones are small in size and can measure below the canoes of trees forming a compacted layer in the rainforest.

Thanks to these studies, it turned out that Amazonian geoglyphs are giant geometric figures: squares, rectangles and circles. The scale of the remains of a culture not yet known is so great that archaeologists from many centres had to be involved in the research to determine what they were actually dealing with.

In the midst of dense forests, international expeditions have stumbled upon the ruins of settlements and large cities, the remains of irrigation systems and farmland. Unexpected and numerous finds indicate that 3–4 thousand years ago, the vast area of today's jungle was densely populated. Little is known about this culture (and perhaps even civilization), but even today geometric geoglyphs delineated by lines of ditches filled with lighter clay are recognized as a characteristic feature of its achievements.

The previously unknown culture of creating geoglyphs has attracted the attention of many scientists from around the world. On February 15, 2015, BBC News reported that research in the Amazon was conducted by a team of scientists from Great Britain. The aim of the project was to determine the scale of activity of the Indian population living in that region in the millennia before the European conquest.

"Until recently, it was thought that the Amazon was "always" inhabited by small groups of hunters and gatherers with minimal environmental impact. It was also believed that the Amazon jungle was an area that had never been affected by changes introduced by high cultures," explained Dr. Jose Iriarte from the University of

Exeter in a statement to BBC News. "New findings indicate, however, that today's rainforest on the border with the Andean zone was inhabited several hundred years ago by numerous, economically complex and hierarchical societies that had a significant impact on environmental changes."

As reported on February 6, 2017 on the University of Exeter's website, a team of Brazilian and British scholars led by Dr. Jennifer Watling of the Sao Paulo University Museum of Archaeology and Ethnography discovered further traces of the ancient Amazon culture, including geoglyphs, which were added as early as 450 at the time.

Geoglyphs are mysterious images created in different eras and in many places around the world. However, the characters from the Brazilian state of Acre are especially original. They have geometric shapes, and they are formed by shallow ditches and low embankmentes located on a huge area of more than 13 thousand square kilometers.

Previously, science assumed that the jungle had been growing in the area for at least a dozen millennia, only hunting and gathering tribes lived in it. However, analyses have shown that the dense forest has been growing in this area for at most two thousand years, and has previously been largely modified by the lost culture. Namely, the Indians for centuries turned the region into a kind of paradise. Skillfully managing resources, they led to the fact that there were mainly useful plants growing there. Some provided food: fruits, rhizomes, leaves. Other wood and fibres for fabrics, cords and braids. However, these were not typical plantations, known today monocultures, which are exposed to destruction by diseases and pests. In fact, in the lost kingdom of the Amazon, an original intermediate form was invented between agriculture and the harvesting of a wild forest fleece.

Another research team discovered geoglyphs in this area as well, as reported in the August 2017 journal American Anthropologist.

Geoglyphs were discovered and tested by an international team led by Dr. Pirjo Kristiina Virtanen who teaches at the University of Helsinki and Dr. Sanna Saunaluoma from the University of Sao Paulo. These symbols are located in the southwestern Amazon in the Brazilian state of Acre and about 500 characters have already been counted. It has been established that these ground symbols took 2,000 years to create, starting around 1050 B.C. until 950 C.E. Geoglyphs are from several to several hundred square meters in size. What sets them apart from other geoglyph complexes is the fact that some marks are outlined with excavations several meters deep. Usually, above-ground carvings are made by exposing the top layer of soil, where the lines are about half a meter deep. The characters from Acre also have simplified geometric shapes. They are mostly circles, ellipses, squares, and octagons.

Though more than 11 centuries have passed since the creation of the last geoglyph, the Indians of this part of the Amazon still regard these marks as sacred. As a result, no fields were created in this area and the symbols survived undamaged. Importantly, these are the patterns that are still present in the art of the Amazon Indians, including applied art as decoration of clothing and ceramics.

According to Doctors Virtanen and Saunaluom, the symbols of Acre were made for ritual purposes. The geometric patterns were shaped like that, because they functioned as gates or roads to the spiritual world and were used by ancient shamans to establish contacts with the spirits of ancestors and deities.

Scientists assume that by using the latest technologies they will solve the mystery of the great earth signs in the Amazon. So far, other research works have identified an element linking the Amazon and Nazi rites. It is one of the great figures carved in the Nazca desert - the one representing a spider. According to the data published on September 11, 2015 on the aforementioned website "Ancient Origins", the analyzes of entomologists showed that although the representation of the spider is only a sketch, its creators marked a characteristic element of the animal's body structure, which indicates that it is an individual of the species Cryptocellus goodnighti.

And here there are as many as two puzzles. First, this characteristic element of the spider's body is the genital organ, which can only be seen with a strong magnifying glass or microscope. Yet such instruments were not available two thousand years ago, that is, during the time of Nazca culture. Secondly, Cryptocellus goodnighti is not found on a dry plateau, but only in one region of the Amazon jungle, more than 1,500 kilometers southeast of the Peruvian desert. So why was the image of this particular spider painted on Nazca? The answer must be only one – in ancient times there were contacts between the Amazon and the Andean area, where the culture of Nazca developed.

The extensive issue of geoglyphs abounds with numerous puzzles. Are the Amazon signs – certainly older than Peruvian and Bolivian – also the world's first geoglyphs?

Did the concept of deuceing great symbols in the ground be invented by the lost culture of the jungle zone, and if not, from whom did it borrow? There are many questions, and few answers, although in the future they may be answered by analysis of geoglyphs discovered in other parts of the world.

Huge symbols carved into the ground, for example, have been found thousands of kilometers north of Nazca in the United States. More than five hundred geoglyphs, known locally as intaglios (or simply "carvings, reliefs"), are engraved in the desert soil along the Colorado River, about 25 kilometers north of Blythe city on the Arizona-California border. Colorado geoglyphs also pose a mystery to science, even though they were discovered almost 80 years ago. Captain George A. Palmer, a US Army pilot, saw them from the air. On November 12, 1931, he made a flight from Hoover Dam base to Los Angeles. As he arrived at the Colorado River, he saw large paintings of people and animals, and geometric compositions carved into the rocky ground. "There are images of snakes and four-legged animals with long tails near the two human figures," Palmer wrote in his memoirs. - And after them, we see a giant or a god who is just quickening his pace to dance in a nearby great circle.

The figure of the giant mentioned by the American pilot is a human figure over 55 meters, and the so-called large circle carved nearby is an oval-like oval with a diameter of 47 meters. These sizes indicate the scale of the undertaking that the former inhabitants of the region faced. Palmer shared the discovery with Arthur Woodward, curator of the History and Anthropology Department at the Los Angeles County Museum of Natural History. Scientists associated with this institution initially assumed that the intaglios had been carved into the ground by the Indians two or three centuries earlier. After a preliminary - and fairly cursory - examination of the symbols, they were no longer interested in the find.

Intaglios has been deteriorating for decades, exposed to the effects of weather and the thoughtlessness of tourists and the ill will of vandals. The first attempt to save the unusual find, comparable to the Nazca geoglyphs, was made in 1957 by the local naturalist De Weese W. Stevens, deputy director of the Palo Verde High School in Blythe. At that time, many signs were photographed and measured, and some were secured with shields.

Another intaglios study - this time overseen by specialists in Indian prehistory - was not conducted until 2009. It has been established that the carvings were not created 200-300 years ago, but more than two thousand years ago. It was also found that the symbols were made using a similar technique as the Nazca geoglyphs, so their creators first delineated the shapes of the figures, and then along the lines removed the top layer of soil up to the level of the light-colored rock.

Scientists disagree as to which people carved the giant symbols along the Colorado River, but the prevailing hypothesis is that they are the work of Hopi Indians, whose descendants still live in that area today. Also, religious studies of symbols indicate that some of these signs can be associated with divine and animal characters found in the mythology of the Hopi people.

The border between Arizona and California is very far from Nazca, Palpa, Sajama, Amazon. Could contacts be made between such distant areas? It is hypothetically possible, but unlikely. But did there have to be a flow of ideas over such great distances? After all, the idea of creating giant engravings could have arisen independently in different parts of the ancient world.

Another example, even more distant from Nazca, is the ground-based images from Great Britain. Scientists assume that the geoglyphs known from the British Isles were created 2–3 thousand years ago, ie they were created more or less at the same time as the engravings from the Americas. They also have a similar manufacturing technique with the American ones, but they differ clearly in style.

British ground-based imagery has been around for a long time, which is perhaps why there is less confusion around them than with the Nazca glyphs. The most famous is the White Horse of Uffington, 111 meters long.

It was probably made 3,000 years ago, in the Bronze Age, and the oldest references to it in written sources date back to the 12th century. Nothing else is known about this geoglyph, and even that it depicts a horse has been questioned. On October 13, 2010, the BBC reported that, according to Olaf Swarbrick, a retired veterinarian, the anatomical features of the White Horse indicate that it is an image of ... a dog.

"The Uffington animal has too long and too thin a body and a tail too long for a horse, even if we assume it's a stylized image," said Swarbrick. - This symbol represents a dog, maybe a hound, which was a species used for hunting a long time ago.

Dr. Keith Blaxhall, an archaeologist from the British scientific association National Trust, under whose care the monument is held, spoke in defense of the traditional "horse interpretation" of the painting from Uffington.

"It's a stylized representation of a horse. Not a literal representation of his character, but a suggestion of what animal it is about," Blaxhall assured. "The shot is also intentional, showing the moment when the horse froze in a gallop."

Olaf Swarbrick's concept was not accepted, although it must be remembered that as a veterinarian with a long experience, he is a specialist in animal anatomy.

The second British geoglyph is the Giant of Cerne Abbas, Dorset. It has the form of a schematic figure of a man - naked, which can be seen from a clearly marked part of the body that scandals some tourists to this day. It's about an erect penis. The figure is 55 meters long and is holding a 37-meter club in his hand.

The third British geoglyph is the 72-meter Tall Man from Wilmington in East Sussex, also known as the Green Man by local residents. His image is carved on the steep slope of Windover Hill. There is a problem with dating this monument, because although it is traditionally assumed that the Tall Man was made in the same era as the White Horse from Uffington, some scholars consider the Wilmington Rite to be a kind of fake. This "scientific opposition" assumes that it arose in the 16th or 17th century, and argues its opinion with the fact that the geoglyph was first mentioned in records from the 18th century. There are also disputes about what the stylized man holds in his hands. Are they two clubs, two spears, or are the farmer's working tools, such as a rake or a scythe?

As you can see, about the three British geoglyphs it is difficult to make indisputable findings. So maybe a little more knowledge about them will bring research on the latest find?

The hitherto unknown British geoglyph was found on September 10, 2004. It resembles a monument from Uffington, because it also presents an outline of the silhouette of a horse (or a hound on the run, as the veterinarian Swarbrick would probably consider). The newly discovered earthen painting is located in a field near the village of Whittlesford near Cambridge.

The Whittlesford horse was spotted by the crew of the avionics and immortalized in the pictures. The geoglyph shows a galloping mount with its head held high. The first analyzes of this mark were carried out by members of the Whittlessford Archives Society, supported by a voluntary team of archaeologists.

Were the few British geoglyphs created independently of the American ones? Logic suggests so, but what if the idea of making geoglyphs was born in one place, and then spread from this "source" to the whole world? This is not a guesswork, because great signs on the face of the earth are found not only in the British Isles or the Americas, but also in other parts of the globe.

Research conducted in recent years in the Middle East has shown that geoglyphs occur over a huge area from Syria, through Jordan to Saudi Arabia. Previously, they were unknown to science as they can only be seen from above.

A series of Middle Eastern discoveries began when, in the summer of 2011, Australian archaeologists (using satellite data collected for new maps) found giant symbols in the shape of spoked circles in the Azraq oasis in Jordan. The details of the unusual find were announced on September 14, 2011 by the "Live Science" scientific website.

Azraq geoglyphs were created at least two thousand years ago, i.e. they come from this period at the latest, as the signs from the Peruvian Nazca desert. Their creators drew symbols on the fields of weathered lava. The diameter of the circular engravings ranges from 25 to 70 meters.

"We came across stone structures in Jordan that are more numerous than the symbols on the Nazca Plateau, cover a much larger area, and may also be older," said Dr. David Kennedy, a lecturer in ancient history at the University of Western Australia.

It is worth noting that the first mention of Azraq glyphs dates back to 1927, when Lieutenant Percy Maitland, a RAF pilot, noticed them from the window of his plane. According to his notes, strange signs were visible from the air "in the land of lava", and when questioned about them Bedouins claimed that they were "works of ancient people". However, the discovery of the British pilot was quickly forgotten.

The signs are indeed difficult to see from the ground level, so Dr. Kennedy's team first determined their exact location based on aerial and satellite photos, and then set off into the field according to GPS indications. Research has led to the discovery of not only numerous geoglyphs made of stones in this region of Jordan. Australians also found many clusters of ruins, such as the remains of stone buildings (now used as animal enclosures), as well as stone burial mounds that were erected in line with these structures.

Kennedy and his associates admitted that they had a problem not only with determining the date of geoglyphs, but also with determining what content (mystical or religious) they expressed and for what purpose they were used by their creators. The structures are assumed to be over two thousand years old, but they may well be much older. This could be indicated by the finds already mentioned - the ruins of settlements and mounds, the age of which was estimated at the 7th millennium BCE. Of course, it would first be necessary to prove that the makers of the geoglyphs were the same as the builders of these Neolithic settlements and mounds.

The earth symbols from Nazca and Sajama are between 1,400 and 2,100 years old. Geoglyphs of the Paracas culture from 2,100 to 2,600 years.

The symbols from the western part of the Amazon seem older than them and they are probably 3-4 thousand years old. British carvings were made about three millennia ago. Dating glyphs from Azraq in Jordan are uncertain - they may be much more than 2-3 millennia old, since - if they are contemporary to the surrounding Neolithic stone structures - they are 9,000 years old. So with each new discovery, we put back the genesis of geoglyphs. So where was the world's first great terrestrial rite created? What people and when came up with this artistic form of supposedly religious significance?

Most of the peoples of modern Europe are Indo-Europeans, who are assumed to come from one area. The birthplace of their ancestors should probably be found in the strip of steppes of central-western Asia, from where they made their way to the west in waves through the so-called Gate of the Peoples, ie the "passage" between the Urals and the Caspian Sea lying on the border of Asia and Europe. It is in the zone of Asian steppes and semi-deserts that special finds were found that connect the geoglyphs of interest to us here with the Aryans (Indoarias), as are the names of those Indo-European peoples who, about 4 thousand years ago, set off from the Central Asian cradle to the south, to the Indus and to India, and also to the south-west, that is to Iran, Syria and Anatolia.

On September 23, 2014, the "Live Science" scientific website presented extraordinary discoveries from northern Kazakhstan, that is, from the area lying in the aforementioned Asian steppe zone. There were found over 50 large geometric geoglyphs in the form of circles, spirals or stylized swastikas. These symbols are difficult to see from the surface of the earth, but they are clearly visible from the air, which is why archaeologists found them in this case while analyzing maps of the Google Earth satellite system. The image documented on the maps from 2013 was so intriguing that a Kazakh-Lithuanian scientific expedition from the Kostanayan State University and Vilnius University headed by Dr. Irina Shevnina and Andrei Logwin set off to the geoglyphs area.

Archaeologists have established that the symbols on the ground are in fact huge, comparable in this respect to the most famous geoglyphs in the world from the Nazca desert, and measure from 90 to 400 meters. Most of the Kazakh symbols were made in a manner typical for geoglyphs, i.e. by removing the top layer of the earth and revealing a lighter layer, clearly distinguishing the lines of the characters.

However, in the case of swastikas, an additional original technique was used: a layer of wood was laid in the shallow recesses of the lines. At this point, one can put forward a hypothesis extending the findings of Kazakh and Lithuanian scholars, because the question arises whether fire was not lit in the lines-ditches wet with wood? After all, the swastika is a symbol of light, sun and eternal life. Perhaps this issue will be clarified by further finds. The more so that archaeologists have unearthed the remains of cult structures and furnaces suggesting that religious rituals were performed in this place. The age of these finds has not yet been established, it has only been assumed that they are at least 3,000 years old. However, the symbolism of geoglyphs indicates their connection with the culture of the Aryans, and in that case it would have to be dated back a thousand or a thousand hundreds of years.

- Today we can only say one thing: geoglyphs were made by ancient people, but by whom and for what purpose, it still remains a mystery - said Andrei Logwin in a statement for BBC News.

Importantly, the area with geoglyphs borders on the Russian-Kazakh border region, where intriguing finds have been made in recent years, including the discovery of the ruins of cities created four thousand years ago. The best-preserved of them, referred to by archaeologists (from the name of a nearby Russian town) as Arkaim, is associated with the already mentioned Indo-European Arias. So, should the geoglyphs of the Asian steppe zone be considered an "imported idea" (for example, from the Middle East), or a local "invention", the work of distant ancestors of today's Indo-Europeans or of the population in general, who lived in the Asian steppe belt in antiquity?

Perhaps this region of the world will turn out to be the cradle of geoglyphs in the course of further research, since the symbol discovered on the slope of the Zjuratkul hill in the Zjuratkul National Park on the southern edge of the Ural range is considered the oldest in the world (because dating is certain here).

In 2012, a Russian expedition encountered him, as reported by the scientific service "Antiquity". The discovery was made by two scientists from Chelyabinsk, a regional city in the Urals: Dr. Stanisław Grigoriew from the Institute of History and Archeology of the Russian Academy of Sciences and Dr. Nikolai Mienshenin from the State Center for Monument Protection. The geoglyph was made on a flat in the slope of the hill, thanks to which it is clearly visible from the top of the mountain - it is the only place, which is difficult to access, from which you can admire the symbol. No wonder that the sign was discovered only on satellite photos, and subsequent findings were made from the plane and powered hang-gliders. Only after precisely locating the geoglyph, Grigoryev and Mienszenin organized an expedition that conducted field research.

The symbol of Zjuratkul is huge. The lines that make up it are almost 600 meters long, and their width is 2 to 4 meters. The geoglyph shows the silhouette of a four-legged, horned animal.

A precise measurement of the entire form is not yet possible, because the area where the creature's back is drawn is covered with dense forest. Image analysis revealed that it is a representation of a deer or elk, animals still found in the region. The shape of the mouth and the type of antler rather indicate a moose.

Unlike most geoglyphs known to science, which were made by removing a layer of soil and revealing a lighter layer below, the lines of the Russian symbol were created by digging shallow trenches and filling them with light-colored clay earth and stones - white quartzite obtained from the top of Zjuratkulu. This filler was poured "downhill", and according to the researchers' findings, the originally convex outlines of the animal protruded above the ground by half a meter. However, the destructive influence of weather conditions and the development of vegetation meant that today the drawing is at ground level.

When was the gigantic image of the moose taken? Scientists have problems with precise dating of the find. The geological system of the region was finally formed 7 to 9 thousand years ago. In turn, palaeobiological and palaeozoological studies of the mountainous area of the southern Urals showed that these areas began to be covered with forest after 2500 BCE.

And since the figure of an animal was drawn on a hillside in a time when there was no forest there, it must have been more than 4,500 years ago. On the basis of their analyses, Russian scientists claim that the Big Moose of Zhuratkul is the oldest geoglyph in the world known to science. Scholars guess that the giant painting had a religious significance and was used to perform rituals related to shamanistic beliefs.

However, shamanistic assumptions alone do not explain the need for geoglyphs, so it's time to reflect on the general issues of great terrestrial signs.

First of all: why did people of that time gather, devote their time and energy to mark out huge symbols in the ground. What were these signs for?

One of the more plausible hypotheses is that the symbols were ceremonial paths. Given that each geoglyph was dedicated to a different deity, followers could perform ritual dances as they moved in procession along lines carved into the topsoil. Each symbol is a different deity, a different rite related to a different season, different religious content.

In such an understanding, geoglyphs should be regarded as calendar signs that helped to organize the spiritual life of a community functioning to the rhythm of rituals. Whether such an interpretation is correct is difficult to judge, as there are also other ideas on this subject.

Now the question of one or more geoglyphic "invented" sites. Do we know where the idea of creating them was born? According to Russian scientists, the oldest geoglyph was created in the Urals (remember, however, that this is only the opinion of the Russians, because the signs from the Amazon may be as old and Jordanian symbols even more ancient). Taking their concept as a starting point, one could hypothesize that the creators of the Great Elk from Zjuratkul belonged to a people who came up with the idea of such a form of worshiping deities, and at the same time developed a technique for performing sacred symbols. So, did the history of geoglyphs begin in the Ural-Kazakh zone? Was the concept of creating great signs on the surface of the earth born in a place where the culture attributed by some archaeologists to the Arians who flourished soon after. And did their descendants, i.e. various Indo-European peoples, propagate the creation of giant rites? However, it is difficult to connect the primitive Indo-Europeans with the geoglyphs of America, and it is also unlikely that the first Indo-European peoples reached the British Isles already in the Neolithic, when the local ground symbols were created.

Or maybe it all started in the Middle East, as exemplified by the glyphs under the Jordanian village of Azraq? The problem is that the dating of the local symbols on the ground is ambiguous - they can be as long as two or nine millennia. As you can see, the mystery of the cradle of geoglyphs is still far from being solved.

And another point. Huge carvings measure from several dozen to several hundred meters, and some even two kilometers, so they belong to the largest artifacts of antiquity in terms of size, so they must have been of great importance for the people who created them. Adding to this the fact that they are discovered on different continents - in the Americas, in Europe, in Asia, the question arises about the possible commonality of their functions. So for what purpose were they performed?

Attempting to answer this question requires much more space than the previous two issues.

The most famous researcher of earth symbols from Nazca, the German woman Maria Reiche, has been dealing with geoglyphs for over 50 years, that is, her entire professional life. She measured and photographed most of the local symbols, but even she did not solve all their mysteries. Currently, geoglyphs are not only dealt with by scientists from large academic centers, but also by numerous amateurs of the mysteries of the past. More discoveries are made, interesting hypotheses are created, and the new findings contribute a lot to scientific knowledge about the signs marked out in the ground.

Nevertheless, the reflections on the symbols from Nazca are repeated. Why? Since they are best known, it is no wonder that most often it is on their basis that scientists try to explain the purpose for which people thousands of years ago undertook to mark great signs in the earth.

It has now been established who, how and when engraved the Nazca glyphs. They were local Indians who, at the turn of the old and new era, set symbols by digging shallow ditches.

However, two issues are unknown. First, how they controlled the precision of the painting if they could not see all their work from a distance. Second, what was the purpose of these activities. Perhaps the answers to these two questions about Nazca will also provide universal answers, concerning all geoglyphs of the world.

The scientists tried to find an answer to the first question using the methods of experimental archeology, thanks to which they recreated the probable method of making earth carvings. It turned out that you can draw a small shape on the ground with a stick and then easily enlarge it by changing the scale.

For this, a few people are enough to use ropes of a certain length and sticks, which are stuck into the ground at places measured with ropes. Then the same group, using shovels and sticks, made lines forming a symbol in the ground, revealing the lighter layer of soil beneath the darker one.

However, not all geoglyphs were made using this method, for example, the Azraq in Jordan and Siberia carvings were made by marking lines with stones. In turn, in the Arkaim area, wooden cladding was also used.

Many researchers, however, assume that in order to precisely delineate the lines that form large earthen images, you need more precision than the above method shows.

According to them, the creators had to see their work from above. Meanwhile, in many places where geoglyphs were created, there are no natural uplift of the terrain from which to observe the progress of works. Nor is it known that people who lived 4-2 thousand years ago used flying apparatuses that allowed man to rise above the earth's surface.

An interesting attempt in this field was made in the late 1970s by a group of American students who went to the Nazca Desert. On site, young Americans made a hot air balloon, the so-called mongolfier, used in Europe since the 18th century when it was invented. Only local materials were used to construct the balloon, both the canopy and the basket. It was assumed that the ancient Indians could heat the air by placing jugs of oil from natural sources in a basket under a balloon. The oil was set on fire and the hot air filled the canopy and the balloon rose.

However, no one has reconstructed these alleged heating devices, so gas cylinders were used instead of jugs of crude oil, as in modern sports or recreation mongolfiers. Therefore, even though the experimental flight of students was successful, the hypothesis that pre-Columbian Indians were using balloons has fallen.

This is not the case with attempts to explain the reasons for creating large symbols. There are many hypotheses on this point - probably too many. According to one concept, geoglyphs are gigantic calendars or astronomical observatories. According to another - tribal symbols, and according to another - places of worship.

Another interpretation of the creation of these signs was presented in 2006 by a team of researchers from the University of Massachusetts. Dr. David Johnson, hydrologist Stephen B. Mabee and archaeologist Donald Proulx assumed that the meaning of the long single lines between symbols was explained by hydrology.

They assumed that the Nazca geoglyphs were routes leading to dry springs today. American scientists admit, however, that they do not know whether all the earth drawings in the desert played such a role, because there are many more of them than the ancient water sources discovered by geologists and hydrologists in this region.

Yet another hypothesis on the role of Nazca geoglyphs was presented in April 2015 by the aforementioned doctor Masato Sakai from Yamagata University, who also discovered previously unknown symbols carved in the Peruvian desert. Sakai's concept was announced on May 1, 2015 by the "Live Science" scientific website.

A Japanese scholar believes that some of the Nazca geoglyphs, namely four sets of figures, pointed the way to a nearby pre-Inca temple. It was the great Cahuachi sacred complex to which the local people made pilgrimages and brought offerings. Sakai also developed a theory regarding all Nazca signs - according to him, they were created in two periods, and were made by two different peoples living in this area several hundred years apart. At this conclusion, the Japanese scientist led some differences in the treatment of signs: at the intersection of some lines, archaeologists found vessel shells that had been broken there, perhaps for ritual purposes.

In turn, American archaeologists from the Pennsylvania State University presented in May 2003 a hypothesis on the role of earth carvings from Sajama in Bolivia. They concluded that the geoglyphs there were closely related to the ancient buildings and tombs in that area. According to scientists, the symbols from Sajama have astronomical significance and reflect, among other things, the constellations of stars, including the Pleiades, as well as the positions of the solstices. Scientists assumed that the earthly book of Sajama was created by a culture that developed between 200 BCE. and 650 CE The gigantic geoglyph complex is intended to reflect the cosmogony of this pre-Columbian culture.

As already mentioned, too many hypotheses have been made about the reasons for creating geoglyphs. Some of them even assume the interference of extraterrestrial factors. For example, the co-creator of the palaeo-astronautics doctrine, Erich von Däniken, believes that carvings such as those from Nazca were made by people in memory of the landing of aliens.

The assumptions of a group of unconventional researchers from Russia are similar to the paleo-astronautical concept. They were presented in December 2012 by some websites devoted to ufology and paranormal phenomena. According to these data, a group of about 50 scientists of various specialties deciphered the meaning of large symbols and numerous puzzling objects, ranging from Nazca geoglyphs to pictograms imprinted in grain and megalithic structures such as Stonehenge. The works of the Russian team are endorsed by a very interesting person, Marina Popowicz - a retired military aviation colonel, and at the same time a UFO researcher and author of books on this subject.

As a test pilot, she set 107 aviation world records, was repeatedly awarded for her services to the country, and in 2007 the

Order of Courage was personally presented to her by Vladimir Putin. Now he supports a team that has reportedly worked for several years to decode information hidden in big characters.

But who encrypted this data in these objects? Apparently some extraterrestrial intelligence. Who made them? Same strength. And what do messages encoded in enormous symbols tell us?

Well, the hidden message heralds a specific end of the world, which is to come after the abrupt change of magnetic poles. As a result of this process, people will reach a different dimension of consciousness. According to Russian enthusiasts, this process was already initiated in 2014.

No matter how fantastic some of the geoglyphic hypotheses are, the fact remains that the purpose and design of these giant signs still remain a mystery. Importantly, the symbols of Nazca and other places are not visible from the ground, and even if there are hills nearby, it is rare to see the symbols in all their glory from their top. Usually, only an observer located high above the ground, for example in an airplane, is able to see them.

The issue seems extremely important, but so far it has not been possible to clarify who was actually viewing the geoglyphs, since humans could not. Gods?

Let's go back to the reasons why people flocked, put their time and energy to mark out huge symbols in the ground. What were these signs for? The most likely hypothesis is that they were ceremonial paths. If we assumed that each sign was dedicated to a specific deity, then the followers could perform ritual dances, creating a procession and moving along lines carved in the ground.

Every symbol is a different divine being and a different rite to honor that heavenly being. From this perspective, geoglyphs constituted a calendar because they helped organize the life of the community, setting its functioning to the rhythm of seasonal rituals.

But is this the definitive explanation of why geoglyphs were created?

Does this hypothesis really reveal the secret of gigantic artifacts, symbols carved into the ground by different peoples in different epochs and on several continents?

Portrait of the god Naylamp

In early 2010, a team of archaeologists from the Peruvian Regional Archaeological Museum of Hans Brüning in Lambayeque, led by Dr. Carlos Wester La Torre, discovered the ruins of the Naylampa Temple. Who was Naylamp? A special figure for the realities of pre-Columbian America, because according to the myths and legends of the Peruvian Indians, he was to create a powerful state of the Chim people, existing from the 9th to the 15th century.

Archaeologists from the Brüning Museum conducted research at the sites of Chotuna and Chornancap, which once together formed a city-temple complex with an area of 20 hectares, located in the area of today's city of Lambayeque in north-western Peru. Chotuna-Chornancap was one of the largest centers of the pre-Inca epoch, and in the times of the Chimu culture, it was second only to the capital of the Chim state - the city of Khan Khan. Near the pyramid in Chotun, Dr. La Torre came across the remains of a 2,500-square-meter temple built of adobe, i.e. sun-dried bricks.

Inside there was a well-preserved stone throne on a pedestal shaped in the shape of the so-called Andean cross and religious paintings on the wall, including a portrait of Naylamp.

- The temple is directly related to the figure of Naylamp, the legendary creator of a culture that once developed in the Lambayeque valleys, and called the Chimu culture - said La Torre. "Interestingly, according to Chim tradition, Naylamp had special physical characteristics: he was very tall with light hair and blue eyes.

The myths and legends of the Chim were written by a Spanish chronicler, priest Miguel Cabello de Balboa, who lived in the years 1535–1608.

According to these accounts, Naylamp came to the Lambayeque Valley at the head of a group of warriors. They appeared not on foot, but came from across the sea in large boats. When could this event happen, if we accept the historicity of this Indian dynastic legend? Probably in the middle of the 9th century CE, i.e. several dozen years after the collapse of the Mochica culture, whose population created a state ruling over the coast of northwestern Peru from the 2nd to the end of the 8th century. After the fall of the Mochiki rulers, political chaos engulfed the local tribes. It was only Naylamp who was to collect them under his leadership and build the cities discovered today by archaeologists and initiate the development of a new state, the area of which is defined by the archaeological term of the Chimu culture (so the population basis of the new state were the descendants of the former subjects of the kings of Mochiki). And perhaps there is a grain of truth in the Indian account of Naylamp's arrival, as the temple devoted to him at Chotuna-Chornancap dates back to the second half of the ninth century, which would be in line with the era of unification of the indigenous Indian tribes after the turmoil. So is the first king of the Chimu state not only a legendary figure?

You should start with his image. As I mentioned, Native American tales recorded by a Spanish priest claimed that Naylamp was much taller and stronger than the local people, with light hair and blue eyes.

Interestingly, it is similarly shown in a painting from a temple discovered by archaeologists. The argument in the form of a monument is strong, but even it does not have to testify to the historicity of the figure of the first king of the Chim.

In fact, the characteristics of this character known from legends and iconic painting could only be treated as elements of a myth, if not for other finds from the New World.

These are archaeological and anthropological discoveries that indicate intercontinental contacts between medieval European sailors and the Indians. Could Naylamp then be a Viking, a Scandinavian warrior-sailor who reached the South of America over a thousand years ago and influenced the lives of the local indigenous people?

In the mythology of pre-Columbian peoples of Peru, there is another character depicted in a similar way to Naylamp, i.e. as a bearded, light-skinned man. This is the Inca god Viracocha. It should be emphasized that his name in the Native American languages of Quechua and Aymara means Man of Sea Foam, so he is also unequivocally related to the sea. Apart from descriptions of various supernatural features and the history of Viracocha, myths say that he liberated man from the state of primordial wildness, that he was the creator of culture, law and social hierarchy, that he introduced the cult of gods, that he taught his subjects to cultivate the land. In the end, however, discouraged by the pettiness and other negative qualities of humans, he sailed - using his cloak like a boat - westward, to the land of the dead. The followers of Viracocha hoped that their god would come back one day - also by sea. When the Spanish conquistador Francisco Pizarro traveled to Peru in 1531, one of the basic factors that allowed Europeans to conquer the multi-million dollar Inca Empire was the cult of Viracocha. The Indians linked the arrival of the bearded Spaniards with tales of the return of the good god. By the time they discovered that there was little divinity and goodness in the conquistadors, the empire had collapsed into ruins.

The Inca state was founded in the central Andes in the mid-12th century, and in the course of its conquests it grew into an empire controlling most of the west coast of South America. It was called Tahuantinsuyu, meaning 'The Four Parts,' because it was administratively divided into four great areas. At the beginning of the 16th century, it covered an area of almost two million square kilometers and had up to 15 million inhabitants. The capital city was the city of Cuzco, the principal deity of Sun-Inti, and the official language of Quechua.

In turn, the cult of Virakochy was professed by the Aymara Indians, whom the Incas incorporated into their kingdom in the fourteenth century. It was then that Virakocza took a prominent place in the Inca pantheon. An interesting trace of possible connections between some alleged overseas visitors and the Indians of Peru documents the story of one of the semi-legendary Inca rulers. The eighth Inca Sapa, the king of Inca presumably reigning in the years 1410–1438, was initially called Cusi, and after taking power he assumed the name of Virakocha.

According to the dynastic record, Cusi was the eldest son of King Yahuar Huacac and his main wife, Queen Ipaucoma, and thus the rightful heir to the throne. However, as he grew older, a beard grew on his face, which was unheard of for the Inca, as it is for most

Indian peoples. Legend has it that members of the neighboring Aymara tribe had facial hair, so the Inca priests ruled that either the queen was guilty of treason, or that the young man's stubble was a punishment sent by the god of thunder and storms to the king for not following the ritual fasting during the birth of his son.

King Yahuar Huacac decided to banish Cusi and deprive him of his right to the throne. After three years, however, the prince returned to the capital, claiming that he had experienced a revelation during the banishment: the god Virakocha was speaking to him and warned that one of the tribes subordinate to King Huacac would stir up a revolt. At first, the young man was not believed, but when the Chanca tribe reached for their weapons and headed for Cuzco, Huacac fled the capital. The bearded prince Cusi, at the head of his faithful Aymar troops, defeated the rebels, and the Inca aristocracy decided that he deserved power. Then the new king took the name of his divine guardian, that is, Virakocha.

This ruler later became famous for conquering the lands of today's northern Chile and north-west Argentina, carrying out large-scale irrigation and construction works, as well as his work, because he was also a poet - the author of religious songs.

Could the unusual facial hair of King Cusi and his close ties to the Ajmarians, who were attributed to the fords by Indian tales, suggest that in the Middle Ages this people came into contact with some bearded visitors from overseas, for example with Scandinavian Vikings? In antiquity and the Middle Ages, the Aymara people inhabited parts of Bolivia, Peru and Chile, and their main center was the great city of Tiwanaku, which from the 6th century BC. until the 12th century AD it was the center of a highly developed culture covering the region around Lake Titicaca. In the period from the eighth to twelfth centuries, the Tiahuanaco empire covered most of the lands that later became part of the Inca state. It was not until the 14th century that the Aymara people were conquered by the Incas. Could the Aymars, then, in the period when they were rulers of a great state, kept in touch with some "bearded" strangers from overseas, which resulted in the creation of the Virakocha cult? If so, we will see right away that neither they nor the Chimas discussed earlier were exceptions in this regard on the American continent.

Let's move far north, to Mexico, where the Aztecs, before the Spanish conquest, led the strongest federation of states in the Mesoamerican area, with the capital in Tenochtitlan, where the city of Mexico is today. The Aztecs came to central Mexico in the 12th century, and from the beginning of the 14th century they began building an empire in which the cult of Quetzalcoatl played an important role. This god was previously worshiped by the Toltec people, who built a high culture in Mesoamerica long before the Aztecs. Quetzalcoatl (like his predecessors of the gods of the various peoples of the Valley of Mexico before him) was imagined as a white-skinned man with pale eyes. He, too, like Aymara-Inca Virakocha, was a god-civilizer: he looked after people, especially craftsmen, gave them knowledge about the calendar, heaven and the underworld. He was believed to have come from a distant overseas land to the east, to which he had returned after completing his mission. Since then, the Toltecs, Mystecs, Mayans and many other peoples of the Mexican area - and consequently the last of the area's rulers, the Aztecs - have been waiting for his miraculous return.

The Mesoamerican Quetzalcoatl (occurring under many names in different peoples, for example, the Maya took its cult from the Toltecs, but worshiped as Kukulkana) had another important trait that connected it with the Peruvian Viracocha. It had the shape of a snake covered with feathers - we will return to this important feature soon. When the conquistadors led by Hernan Cortez invaded the Aztec empire in 1519 and imprisoned their king Montezuma, the victory of the Spaniards - as in Peru - was largely possible thanks to the belief of the followers of Quetzalcoatl that the Europeans were the emissaries of a returning deity.

The Quetzalcoatl Myth made the conquest easier for the Spaniards, and it brought terrible results to the Indians. It was influenced by the physical characteristics of Europeans and technologies used by newcomers unknown in pre-Columbian America. Had some sailors from Europe crossed the ocean several hundred years earlier, precisely in the era of the alleged deity's activity, giving rise to a myth? Assuming that this was the case, the Scandinavian sailor-warriors are most suspected of it.

For several hundred years, the Vikings traversed the seas on drakkar - ships over 20 meters long, made of oak and pine wood and decorated with a dragon head carving on the bow. Viking long boats from the early 8th century to the 11th century reached France, the British Isles, Orkney Islands, the Faroe Islands, Iceland, Greenland, Spitsbergen, North America, Italy, the Black Sea and the Caspian Sea. Importantly, the Vikings were not only robbers - they were also engaged in trade, and in the late period they conquered the attacked lands. In the north of France, they created the principality of Normandy, in the British Isles they built a network of small kingdoms, and in Eastern Europe they created a vast state - Ruthenia. These excellent sailors also reached North America.

The fact of crossing from Europe to America on small vessels has become myths in the public consciousness, but it turns out that it was much easier than scientists used to think. Practical evidence has been provided by various experimental expeditions, such as the 2005–06 cruise of two Canadians, 34-year-old Colin Augus and his fiancée, 31-year-old Julie Wafaei.

The pair left Portugal on September 22, 2005, and reached Puerto Limon, Costa Rica, in late February 2006. Augus and Wafaei crossed the Atlantic in a rowing boat 7.1 meters long and 1.9 meters wide. The boat weighed 750 kilograms and was equipped with two cabins. Of the modern instruments, only a seawater desalination device was mounted, because the unit had too little displacement to accommodate an adequate amount of fresh water.

Canadians paddled 18 hours a day, struggling against storms, huge waves and scorching sun. Their feat proves that transcontinental ocean expeditions can be made even in very small vessels. All the more, brave Vikings could set off on such voyages.

Another expedition worth mentioning traveled to America in the summer of 2001 on a faithful replica of the drakkar. The ship, named "Islendingur", crossed the Atlantic from Iceland to Newfoundland, covering over 4,000 kilometers of oceanic expanse. The builder and captain of the replica of the Viking ship, Gunnar Eggertsson, claimed it came in a straight line from the famous Leif Erikson. The replica of the Drakkar was constructed, like the original Viking ships, from oak and pine wood. The unit was 22.5 meters long and had a displacement of 80 tons.

In Viking times, the crew of a ship of this size would be 70 sailor-warriors, working two shifts at 32 oars. According to today's standards, such a ship is a shell, but it was precisely similar-sized Drakkars from Europe to America. The ancient Scandinavian sagas say that the expedition, led by the alleged mantis of Captain Eggertsson, the Viking commander Leif Erikson of Greenland, reached Vinland around 1000 AD.

Scientists identify this land with the north-eastern coast of North America. In recent years, many traces have been discovered indicating that the Leif sagas describe true events and that Erikson's expedition was not the only Vikings' foray into the New World. For example, on the coast of Newfoundland in the village of L'Anse aux Meadows, in 1960, archaeologists discovered the remains of wood and earth buildings and objects characteristic of the Viking culture from the times of Leif Erikson.

Another confirmation of the presence of medieval Scandinavians in America was provided by genetics. Two scientists from the University of Iceland in Reykjavik, anthropologist Gisli Palsson and biologist Agnar Helgason, used this modern scientific method to explain whether there were relationships between the Vikings and the Inuit, or Eskimos from the north of the American continent, close enough to produce offspring. Researchers from Reykjavik were prompted to undertake such searches by rumors about the existence of people with European facial features among Inuit.

The Eskimos show mostly Mongoloid features, they are short, stocky, with dark hair and eyes. However, according to many accounts, Inuits with light hair and blue eyes were still encountered at the beginning of the 20th century. According to Palsson and Helgason, these mysterious white-skinned Eskimos were descendants of Vikings who wandered north of the New World between the 9th and 11th centuries. Icelandic sagas as well as Inuit legends tell about such contacts (both about wars being waged and pacts sealed by marriage). In the tradition of both cultures, there are permanent traces of meeting strangers. In addition, testimonies of meetings in the far north with Scandinavian-type people living in the Eskimo way are found in the records of many whaling ship captains that venture into the waters of the Canadian Arctic. This is also confirmed by the notes of John Franklin and Vilhjalmur Stefansson, researchers of the land of ice cream at the turn of the 19th and 20th centuries.

- This could be evidence of more than just economic exchange. So the Inuit and Vikings met not only in western Greenland, Palsson said. - Such findings shed new light on the history of the Inuit.

Icelandic scientists began their research in late 2002. By the fall of the following year, they had 350 genetic samples of modern Inuit living on the shores of Cambridge Bay in northern Canada and on the west coast of Greenland. This material was compared with the characteristics of the DNA of medieval Scandinavians, previously established by other researchers. The results showed that there were in fact close contacts between Eskimos and Vikings.

"The discovery that Inuit and Vikings intermingled about a thousand years ago changes our perception of the mobility of people of that time," Palsson said. - Archeology and anthropology increasingly show that areas previously considered natural barriers were in fact routes of migration.

An artifact may indicate that the Vikings traveled even farther into the New World a thousand years ago - on the prairies of North America, into what is now Oklahoma. Evidence of not only expeditions, but also the settlement of Scandinavian sailors-warriors in the hinterland is to be a boulder with an inscription made of runes on it. The flat boulder called "Heavener Runestone" is over 3.5 meters high and 3 meters wide, and is located near the city of Heavener.

According to some researchers, this runestone is supposed to confirm that the Scandinavians, having crossed the Atlantic in their long boats, reached the eastern coast of the continent, then sailed deep into the continent on the great rivers of Mississippi and Arkansas, and from there by land to Oklahoma.

They were to build a settlement and farm fields in this area. However, the only trace of this today is a stone covered with runes.

What does the alleged Scandinavian Oklahoma inscription convey? According to some researchers, the date November 11, 1012 was engraved on it. Others claim that the runes have written the name of a place identified as "Glome Valley." It is not known, however, whether it concerns the American Viking settlement or the region from which they came from the Old World.

Modern European settlers stumbled upon this runestone in the mid-nineteenth century, when the region was opened to colonists following the forced relocation of indigenous Indians in 1838. The white visitors considered the mysterious inscription to be magical characters engraved by the Indians, because the runes did not resemble the Latin letters they knew. It was not until 1920 that a resident of Heavener copied the characters and sent them to the Smithsonian Institute in Washington. The first analysis of the researchers was unequivocal: there was an old Norwegian inscription on the boulder. At the same time, scientists stipulated that this finding is pointless, because it is well known that medieval Scandinavians did not live in America. Due to the impossibility of establishing the date of the inscription, it was assumed that the runes had been carved recently by a settler from Norway who was a fan of the Scandinavian tradition.

Heavener Runestone was forgotten for decades. It was only thanks to the efforts of the local teacher, Gloria Steward Farley, that the stone became a local curiosity.

Farley studied the inscription and published her findings about it. According to her, it was created between 600 and 900.

Above all, however, the teacher created a state park with a runestone that has been protected since then as its central point.

Gloria Farley's actions did not end there. She claimed to have found four more places in Oklahoma marked with Viking runes. Officially, however, the age and authenticity of these artifacts have not been verified.

To this day, the Haevener runestone remains a mystery, although it must be made clear that it is not recognized by science as reliable evidence of the presence of Vikings in the American hinterland.

Did the Scandinavian sailors leave behind other - and researchers not objecting - traces in the North American highlands? Once it was suspected that they were responsible for the fall of Cahokia, the largest city of the pre-Columbian era in the area of today's United States (Cahokia was situated in the Mississippi oxbow lake near today's city of St. Louis). On May 19, 2015, the National Geographic website reported, a team of experts led by Drs Samuel Munoz and Jack Williams, geographers at Wisconsin State University in Madison, surveyed the Cahokia area. It was established that the Indian city reached its peak of development in the 11th and 12th centuries CE. It then became the main political and cultural center of a large cultural area, which was probably a unified state headed by a chief-king-priest. Cahokia had a population of tens of thousands. The central position in the city was occupied by temples and "chieftains' houses" on top of great earth pyramids, and large wooden public buildings surrounded by hundreds of residential houses.

Cahokia resembled the metropolises of the pre-Columbian cultures of Central and South America, but was an exception in the northern continent.

Analyzes by the Monoz and Williams team showed that the city ceased to exist in the second half of the 13th century. What caused his fall? Scientists have considered several explanations: prolonged drought, catastrophic floods, over-exploitation of resources in the region, and finally the effects of armed conflict.

However, no evidence has been obtained clearly indicating the direct cause of the fall of Cahokia. The city was defended by a system of palisades and earth fortifications, and besides, it was so populous that in order to get it, the attackers would have to have a really large army. And if the Vikings did make it this far inland, they formed small teams rather than contingents of a few thousand men. The most important argument, however, is dating: in the second half of the 13th century, i.e. during the fall of Cahokia, the Scandinavians had not set out on Atlantic Viking expeditions for a long time (i.e. for almost two centuries).

So it seems that the Vikings had nothing to do with the destruction of the only great Indian city in North America. However, I am about to present other scientifically confirmed finds about contacts between Scandinavians and Indians.

It is true that they do not relate to the territory of the present USA, but they will allow us to return to the hypothesis presented at the beginning of the article that King Naylamp, immortalized in the temple portrait by the Chimu Indians, was a Scandinavian warrior.

In January 2006, the scientific journal American Journal of Physical Anthropology published the results of the research by Dr. Caroline Arcini, an anthropologist at the Swedish National Heritage Institute in Lund.

She examined the thousand-year-old remains of Vikings discovered in recent years in Denmark and Sweden. Arcini reported that on the teeth of 25 young (and therefore probably killed) warriors she found unusual decorations in the form of characteristic parallel grooves. It turns out that this type of tooth decoration has so far been known only from pre-Columbian finds from Mexico and the southern US states.

- These remains are dated from 800 to 1050 - said the Swedish anthropologist in a statement to the media. - The grooves carved on the teeth of the examined skulls are made very precisely by a person with great dexterity and experience.

Dr. Arcini assumed that such distinctive dentition marks could be considered evidence of Viking transatlantic expeditions and cultural exchanges between Scandinavia and Mesoamerica in the early Middle Ages. The early Scandinavians either learned this method of decorating teeth in Mexico, or even the decorations were made in America, which would mean that those warriors buried in Scandinavia, whose remains were examined by Arcini, were personally in the New World.

The meaning of the patterns on the teeth is not clear. Scientists speculate that it could mean belonging to a specific professional group, for example, sailors who set out on the farthest, most dangerous voyages. But they might just be ornaments that the warriors in the new land liked and began to copy them.

The contacts of the Vikings with America are also indicated by a sensational discovery from the Norwegian city of Sarpsborg, located less than 75 kilometers from Oslo. The information about this find was reported on June 26, 2007 by the Norwegian daily "Aftenposten". Archaeological research of the ruins of the church of St. Nicholas in Sarpsborg was launched in 2005. This temple was erected at the very beginning of the city's existence, i.e. shortly after 1016, during the reign of the Norwegian king Olaf, later recognized as a saint. Under the church's floor, archaeologists, led by Dr. Mona Beata Buckholm from the Borgarsyssel Museum, discovered the graves of three people: two men and a child. It turned out that the burials date back to 980–1015, that is, from the period when a Christian temple had not yet been erected there.

Examination of human remains led to a surprising discovery: one of the men, 45–55 years old at the time of his death, was an Indian from Peru. This is indicated by the characteristic anatomical features, especially the structure of the skull - there is a bone seam on the occipital bone.

- This feature is inherited only among the descendants of the Peruvian Inca people. This is a real sensation, Buckholm said.

Who was the Indian whose remains were discovered in a grave outside a Norwegian church? A merchant, traveler, slave?

Or maybe a pilgrim wishing to visit the land of the god Virakocha, the Man of the Sea Foam, who once sailed on his coat across the ocean to the west, where the land of the dead lies?

I have presented a lot of discoveries from both the Americas and Scandinavia. Some of these finds may lead us to the conclusion that the Vikings reached not only the shores of North America, but also Central and South America.

Does the mythical image of figures such as Naylamp, the first king of the Chimu people, or the gods Wirakocz and Quetzalcoatl preserve the memory of contacts with fair-skinned European visitors arriving on boats decorated with dragon heads? Had the young Vikings, whose teeth were decorated in the style of Central American Indians, actually returned from the New World? Has a Peruvian Indian buried in a South Norwegian grave really decided to see the land of Viracocha?

There are too many scientifically proven finds to be a coincidence, so perhaps it is worth looking more favorably at the dynastic legends and myths of ancient Indians?

Naylamp might indeed have been the leader of the Viking expedition that reached America, where it played a significant culture-forming role. Is the Scandinavian name and adventures of Naylamp described in a saga that has not survived to our times and therefore this chief is unknown to historians? Or maybe there was no one to tell and write down this saga, because Naylamp had won a new kingdom among the Indians and did not return to Scandinavia? Instead of a saga, he left behind a myth passed on by the Chimu Indians who portrayed their remarkable first ruler on the wall of the temple in Chotuna-Chornancap.

Old China script in the USA

America was discovered by Columbus, but before him the Vikings were reaching the New World. It turns out, however, that the Scandinavian sailors could also have had predecessors. This is to be proved by the characters carved on rocks in various US states, which their discoverer, John Ruskamp, identifies as Old Chinese inscriptions.

John Ruskamp is a retired Illinois chemist who studies alternative American history and has written several books about it. He claims to have found material evidence of the presence of Chinese people on the American continent in antiquity: "in the epoch of Moses and the Trojan War," as he himself puts it. The American service "Mail Online" informed about the research conducted not only by Ruskamp, but also several scientists from recognized research centers. Ruskamp claims to have found inscriptions from the Shang Dynasty, which ruled Northeast China from the 18th or 17th century to the 12th or 11th century B.C.

The rock signs documented by him are located in distant places: in California, Nevada, Arizona and New Mexico, i.e. in the states of the southwestern part of the USA. An amateur researcher identified 84 Chinese inscriptions composed of pictograms, which were already obsolete centuries ago, so their antiquity should not be doubted. The signs, more than 3,000 years old, are to prove that the Chinese not only reached the western shores of North America on their ships, but also wandered far into the continent.

- So far, only half of the symbols on a large boulder in the Petroglyph National Monument near Albuquerque, New Mexico have been identified as Chinese pictograms, but the form of the characters examined and the syntax of their notation are consistent with the rules used in Chinese during the Shang Dynasty - Ruskamp explained.

Other pictographs he found in Grapevine Canyon, Nevada, also seem to date back to the end of the Shang Dynasty, 1300-1100 BC. Importantly, the symbols written in Grapevine Canyon were discontinued in China in the 11th century BC, and the scientists of the Middle Kingdom discovered them and deciphered their meaning only in 1899, so they could not be faked by any of the Chinese emigrants who came in large numbers to the USA in the 19th century

"Ancient Chinese writing proves that Asian sailors not only discovered America, but also influenced the indigenous peoples of North America long before Europe's exploration of the continent," said Ruskamp.

What are the discovered signs talking about? According to the retired chemist, the symbols from the Petroglyph National Monument give details about the voyage, celebrate the Shang Dynasty, and announce the sacrifice for Emperor Da Jai's welfare. In turn, signs from Grapevine Canyon speak of a successfully completed journey to a place called the House of the Sun. There is an unidentified symbol at the end of the text, which may be the author's signature.

Did the alleged contacts with ancient Chinese affect Native American culture? As early as the 1950s, the French sinologist Dr. Jacques Gernet argued that some of the ancient Indian figurines of animals are strikingly similar to Chinese art in the Shang dynasty. Gernet was never proved right, however. On the other hand, bronze artifacts - buckles and a whistle - discovered by archaeologists in the 1990s at Cape Espenberg in Alaska were clearly classified as originating from China, or possibly Korea. However, they found their way to America much later than the symbols found by Ruskamp, around 600 CE. It should be emphasized, however, that the amateur researcher is not alone in the views that newcomers from Asia often came to the American continent, and even influenced the civilization development of the Indians. There is a growing group of scientists who see the possibility of transoceanic contacts.

One of Ruskamp's tenets is an expert on Chinese Neolithic civilization, Dr. David Keightley of the University of California, Berkeley. He even helps a retired chemist decipher the inscriptions on the rocks. Another researcher supporting this concept is Owen Mason, PhD in archeology of Colorado State University.

"We already know there have been contacts between the advanced civilizations of China and Korea and the Native Americans, but we do not know how often they happened and whether they were culturally significant," Mason told Live Science.

Could Ruskamp's markings carved on rocks and a few bronze items found in Alaska change history as we know it? It turns out that not only these finds (causing disputes in the world of science) may indicate that Columbus was not the first to come to the New World from the area of highly developed civilizations of Europe and Asia.

The concept that the Chinese reached America long ago is not new. Even before the aforementioned Genet, in the mid-nineteenth century, the German sinologist Karl Friedrich Neumann connected a deep-sea expedition from around 500 CE (times of the Qi dynasty in the southern part of the Middle Kingdom) with the discovery of America. According to old Chinese documents translated by Neumann, Buddhist monk Hoei Shin obtained the approval and support of the authorities to travel east across the ocean "to the country of Fusang," a great land inhabited by "painted people".

The goal was to spread Buddhism among the barbarians. Sources say the journey was successful, with most missionaries returning home after crossing the Pacific Ocean back and forth.

The cruise was quite a feat because it lasted a year. It wasn't long anyway, considering that the sailors had a distance of about 20,000 kilometers to travel (both ways) - assuming Neumann's view that Fusang is the west coast of North America.

From Chinese sources, we also know the description of the expedition, which is much better documented than Hoei Shin's voyage.

According to the notes from the imperial archives, 71 years before Columbus, a flotilla of huge junks led by Admiral Cheng Ho reached the land identified with America.

In April 2004, the Chinese news service "The Epoch Times" informed about the work on the reconstruction of the flagship of Admiral Cheng Ho, undertaken by the Admiral Czeng Ho Sympathizer Association of Singapore. In this way, Singaporean businessmen of Chinese descent celebrated the 600th anniversary of the admiral's discovery of America.

Cheng Ho (or Zheng He, depending on transcription) was born in 1371 and presumably died in 1433. At the beginning of the 15th century, he was the admiral of the Ming Dynasty Emperor Cheng Zu (also known as Zu Di or Yongle).

In the years 1405–1422 (or 1433), Cheng Ho, at the head of a huge fleet of huge deep-sea ships, led several exploration expeditions. During them, he explored the coasts of India, the Persian Gulf and East Africa, including Madagascar, where Portuguese travelers did not reach until the beginning of the 16th century. The admiral's fleet was also due to sail to Australia three centuries before the official discovery of the continent by British James Cook. Cheng Ho is also credited with a voyage to the New World. Both the expedition to Australia and America are hypotheses put forward by a group of Chinese and Australian historians.

They claim that the admiral's flotilla reached America in 1421, 71 years before the Christopher Columbus expedition.

It should be mentioned here that Cheng Ho's ships were several times larger than those of Columbus. The largest of these junks were multi-masted, with a displacement of up to two thousand tons, 150 meters long, 50 meters wide, and a crew of up to 1,000 people. They were called treasury ships because of the great value of goods that each of them could carry in the hold. These units accounted for a quarter of the mighty armada of 250 ships (the remaining junks were much smaller), which was ordered by Emperor Cheng Zu, an enlightened ruler who focused on the development of trade and knowledge.

The empire-financed expeditions of Admiral Cheng Ho were both about exploring distant coasts and describing them geographically and cartographically for future generations of Chinese sailors, and about developing trade contacts. Although both goals were achieved, the effects of the expeditions were soon wasted because China closed itself to the world and gave up long-range economic and cultural contacts.

Some researchers believe that the discovery expeditions of the Portuguese and Spaniards, which took place at the end of the 15th and the beginning of the 16th centuries, could have taken place thanks to the maps drawn up by Chinese cartographers from the admiral's expedition, copies of which reached Europe. Without their knowledge, neither Christopher Columbus, nor Vasco da Gama, nor Ferdinand Magellan would have been successful.

The Venetian merchant Niccolo da Conti, who had been sailing for many years on one of the ships of the Chinese fleet, was to be brought to Europe by Chinese maps and information helpful in ocean sailing.

One of the evidence supporting this hypothesis is a certain Portuguese map, undoubtedly based on earlier Chinese sources. This map was drawn up half a century before Columbus' expedition, and yet it shows the outline of the mainland of the American continent.

The achievements of the Chinese admiral's expeditions were forgotten for a long time, and a combination of unfortunate circumstances contributed to this. Emperor Cheng Zu moved the capital from Nanjing to Beijing in early 1421. In May of the same year, just a few months after the Cheng Ho fleet set out on the sixth expedition (which, according to supporters of the "American concept", was to reach the shores of the New World along a route along South and West Africa and then across the Atlantic), to the newly erected wooden imperial palace complex called the Forbidden City was struck by lightning. A gigantic fire broke out. The element consumed buildings and most of the archives, and the disaster shook the empire.

The ruler's favorite favorite was lost, and Cheng Zu himself collapsed. These accidents were seen as a sign from heaven, showing the emperor a lack of favor, especially in terms of "sacrilegious" openness to the world, carried out, inter alia, through the cruises of the Cheng Ho fleet. There was a coup and the mandarins took power.

After returning from the expedition, the admiral was removed from performing significant functions, sharing the fate of other protégés of the dethroned ruler. China itself, following the path "indicated by the gods", entered the centuries-long period of closure to the world.

It was not until the 20th century that Chinese scientists recognized the importance of Cheng Ho's discoveries.

Today they are also supported by a group of Australian historians. However, researchers from the US and Great Britain downplay the role of the imperial admiral's expeditions in the field of geographical discoveries and completely deny that he reached Australia and America. Who's right? Can you ignore the Ming vessel finds in America?

Are the even older, characteristic Chinese anchors, discovered on the coasts of South and Central America and Australia, dating back to the beginning of our era, also authentic traces of contacts with Asia? Are they remnants of Chinese trading factories, or, as most scientists suggest, were brought there much later by Portuguese and Spanish sailors as purchased goods in China? Is the 5000-year-old pottery from Ecuador, identical to the Yomon pottery from southern Japan known from the same period, really a trace of migration of the former inhabitants of Kyushu across the ocean?

If it turns out that the documents describing transoceanic voyages from China concern the voyages to America in the fifteenth or even in the fifth century CE, it will destroy the existing scientific dogma about the lack of contact of pre-Columbian cultures with Asian civilizations. Accepting finds of Chinese anchors and Japanese and Chinese pottery in America and Australia will have the same effect. It will be equally important to establish that the rock marks discovered by John Ruskamp are indeed Old Chinese marks, and therefore prove that Asians made cruises to America already in antiquity. However, the acceptance of such theories has not yet been approved by the scientific world. This view could only be changed by new and completely unambiguous discoveries.

Pedra Furada

We already know from previous articles that before it was officially discovered by Columbus, the Vikings, and perhaps also the Japanese and the Chinese (I am deliberately omitting interesting discoveries that may indicate migrations from prehistoric Africa, Palaeolithic Europe, the Mediterranean zone in the megalithic or medieval period) Polynesia).

But what about the first people who settled in the New World? For many decades in science, there has been little violent disagreement on this topic: it was widely accepted that they were newcomers from Asia who, over a dozen (17 to 13) thousand years ago, crossed to the new land via the land bridge, which at the end of the last Ice Age temporarily arose in the Bering Strait.

However, monuments have been found for years which indicate that some people overtook the Asian ancestors of the Indians and came to America several thousand years earlier. In Brazil, rock paintings, which can be as long as 40,000 years old, were discovered in the 1970s. If correct, it would allow the history of America's settlement in a new light.

What is particularly intriguing, according to one theory, the Brazilian paintings are to be the work of a people related to the Australian Aborigines, because they closely resemble the art of the indigenous people of the smallest continent.

The Portuguese sailor Pedro Alvarez Cabral discovered Brazil for Europe on April 22, 1500. In turn, the views adopted in archeology for a long time assume that the oldest traces of the first humans in the South American region date back to 15,000–12,000 years ago (Chile, Brazil, Argentina). However, the rock art works discovered since the early 1970s in Pedra Furada in the Serra da Capivara National Park, located in the highlands of eastern Brazil, prove otherwise.

The dating and interpretation of this particular collection of 350 different scenes captured on stone walls is controversial in the academic world. First of all, the indicated age of these mysterious works of art is in stark contrast to the scientific assumption about the time of human settlement of the Americas. According to a group of Brazilian scientists led by the discoverer and principal researcher of Pedro Furada's paintings, Dr. Niede Guidon, an archaeologist associated with the Sorbonne in Paris, most of these rock paintings were created much earlier than the timing of people's arrival in America. Dated C14 radioactive carbon in 1986 showed that the Pedra Furada artifacts were created in the time interval between 40 and 30 thousand years ago.

Currently, Guidon claims that the oldest of the paintings were created 48,000 years ago, and the youngest - 32 millennia ago. So who, if officially no man at that time lived in Brazil, could have been the creator of this art gallery?

It is not only the age of these monuments that is surprising. After some time, the riddle of Pedro Furada was reminded of the world by another hypothesis of Brazilian researchers, who pointed out the remarkable similarity of the local paintings to the ancient works of Australian Aboriginal art. Importantly, works created in the same era as the paintings from Pedro Furada, and created in Australia already in the first period of its settlement, that is between 60 and 30 thousand years ago.

Opponents of Niede Guidon's findings and the hypothesis based on them emphasize that the Aborigines are nomadic peoples, which although they have a developed spiritual culture, but their material culture is very primitive. It is true that before the Australian continent was conquered by Europeans, some of the Aboriginal tribes living on the coast were engaged in fishing, but there is no evidence that they were able to build boats that would cross the ocean. In addition, critics of the very early dating of the paintings from Pedro Furada emphasize that the oldest stone tools found in that region date back less than 13 millennia, and thus agree with the classic dating of the time of human arrival to South America, and at the same time deny the possibility of moving back the century of the paintings by several dozen thousand years.

However, supporters of the concept of the settlement of South America by the ancient Aborigines also have arguments. First, they show the striking resemblance of the Pedro Furada paintings to the Aboriginal rock paintings of the Kakadu National Park, located on the north coast of Australia, near the city of Darwin. Research has shown that these works were created 40,000 years ago, which is the same era that Guidon indicates as the time when Brazilian paintings were made.

Secondly, the researchers refer to the analysis of charcoals found near a Brazilian cave in the 1980s.

They showed that they date back even 60 millennia. However, it is not certain that the charred wood comes from a fire started by the creators of rock works, and is not the result of a natural forest fire (these doubts were rightly raised in the 1990s). So did the rock paintings from Pedro Furada actually started to be created 40 (or even 50 or 60) thousand years ago, and are they the work of Aborigines who came from Australia? It turns out that further discoveries support the hypothesis presented by Dr. Guidon.

According to recent findings by Australian researchers, tens of thousands of years ago some Aboriginal tribes created cultures more advanced materially than those that have survived to our times. Today's popular view of the primitivism of the indigenous peoples of Australia is far from the truth, because it is important to remember the extermination that these peoples succumbed in the 19th century as a result of the British expansion.

The aborigines were decimated by diseases brought from Europe, and the colonists literally hunted them. Less than a century passed, the number of Aborigines plummeted from 400–600 to 60,000.

The escaped tribes found refuge deep in the continent, in the country's most inhospitable desert regions.

With the extermination, most of the most advanced tribes have also disappeared, while archaeological research shows that in the past not all natives were nomads living in the barren wastes. There were also numerous tribes of fishermen who formed societies with a much more economically developed way of life. Moreover, monuments indicating the cultural advancement of the Aborigines in certain areas of life, many millennia ago, were also found.

In February 2006, the ABC news service reported the discovery of remains of a stone settlement in Australia. These structures were built 30,000 years ago and are located in Mount Eccles National Park in Victoria, on the southern coast of the continent. It is the site of the indigenous Tyrendarra Conservation Area.

The traces of the settlement were found accidentally after a fire consumed 240 hectares of bush, burning 90 percent of the vegetation.

In this area, inaccessible before the disaster, the ruins of a dozen or so long and 5-meter-wide stone houses were discovered. They were situated on the hills above the bed of streams that used to flow there. The remains of artificial water tanks and stone tools were also found.

"The traces found show that this place has been inhabited for a long time by an organized and settled community," said Dr. Damein Bell, head of the organization managing the territory of Tyrendarra. - The life of the inhabitants of the settlement revolved around the water system, which could be a source of food for them and serve as a trade route.

Thanks to this type of finds, it is known that at least some of the original inhabitants of Australia were culturally advanced - they gave up nomadic life in favor of permanent settlement and were able to build stone houses.

That is, it was more advanced than the peoples who inhabited Europe in the same era. In the light of archaeological knowledge, the settlement of Tyrendarra is the oldest stone settlement in the world, and what is important in relation to the paintings from Pedro Furada, the site of Tyrendar comes from an equally distant time.

It doesn't stop there. On February 4, 2011, the British Daily Mail reported the discovery of an Aboriginal stone structure that served as an astronomical observatory. The remains of this structure - bearing many megalithic features - were found near Mount Rothwell, 50 kilometers west of Melbourne.

"This stone structure made it possible to accurately determine the apparent course of the sun in the sky," said Dr. Ray Norris, an astrophysicist at the Canberra-based Australia-based Organization for Scientific and Industrial Research. - The implementation of this object required very precise measurements and knowledge of the basics of astronomical knowledge.

According to the findings of a team of experts led by Norris, the Mount Rothwell Observatory was built between 20 and 10,000 years ago.

The above-mentioned two examples of discoveries constitute premises for a hypothesis that in the distant past, the Aborigines were culturally developed enough to have the necessary skills, including astronomical knowledge, thanks to which they could undertake long expeditions in open waters, and thus also cross the ocean.

In addition to the hypotheses based on archaeological finds, other types of evidence have emerged. They were provided by geological, anthropological and DNA analyzes.

In July 2005, a group of researchers from Mexico and Great Britain led by Dr. Silvia Gonzales, an anthropologist at John Moores University in Liverpool, showed that people did live on the American continent at least 40,000 years ago, at the time of the Pedro Furada paintings. Dr. Gonzalez's team discovered near the city of Puebla, a hundred kilometers south of Mexico City, a multitude of human footprints preserved in petrified volcanic dust.

The eruption of the volcano there was set at 40,000 years ago. As a result, the entire area was covered with a layer of dust.

The volcanic ash quickly binds to a permanent coating, so since footprints have become permanent in it, people must have passed it shortly after the eruption. However, whether they were visitors from Australia - and such a conclusion would greatly strengthen the hypothesis about the "Aboriginal" origin of the paintings from Pedro Furada - unfortunately it is not known. After all, they could have come from another part of the world, such as Asia.

However, this is not the end of the Mexican thread, as the same team of researchers found in 2004 that 13,000-year-old skulls, found in the late nineteenth century near Mexico City (stored in the National Anthropological Museum), have features characteristic of the anthropological type of Australian Aborigines.

Another solid evidence in favor of the ancient presence of Aborigines in America was provided by experts on a team led by Dr. Pontus Skoglund, a geneticist from the Harvard University Medical Department in Boston. Their findings were announced on July 21, 2015 by the Live Science website.

I have already written that earlier genetic studies indicated that the Indians descended from the ancient Asian population, whose representatives reached America several thousand years ago, at the end of the Ice Age. Indeed, to this day South and Central America is dominated by the anthropological type derived from the Mongoloid inhabitants of Asia. However, thanks to Skoglund's research, it turned out that the Brazilian Indian genotype contains a specific memento from much earlier times.

The Harvard team conducted DNA analyzes of representatives of 30 indigenous groups of Central and South America, and then compared the results with the genetic material of people from 197 populations around the world. The comparison showed that even today some Indian tribes living in the Amazon jungle show characteristics closer to the peoples of the Andaman Islands, Papuans of New Guinea and Australian Aborigines than to modern Asians and Indians. The most visible is the link between these Brazilian Indians and the Aborigines.

"The genetic studies so far have indicated that Native Americans are descended from a single source population that came from Asia," explains Skoglund. - It turns out, however, that the Indians have more ancestors, and the settlement of the Americas was a very complicated process.

This is evidenced by the fact that about two percent of the population of modern Amazonians is descended from the People Y. "People Y" is a term scientists have applied to the Amazon Indians of the "Aboriginal lineage". The discriminant "Y" was taken from the word ypykuera, which in the language of the Brazilian Tupi tribe means ancestor, and it was among the Tupis that the most genetic traces leading to Australia were found.

"We know too little to say anything about the history of this lineage, but we believe that Aboriginal DNA came to America earlier than Asian DNA," Skoglund said in a statement for Science. In fact, this discovery raises more questions than it answers.

One thing is for sure. Australian newcomers left their genetic signature in the Amazon. This may have happened at the same time that the peoples of Siberia crossed the Bering Strait to North America several thousand years ago, but presumably happened much earlier, perhaps even 40 millennia ago. Perhaps the same Aboriginal explorers left their footprints in the petrified volcanic dust beneath the Mexican city of Puebla, and then continued south to the Amazon, finally reaching the highlands of eastern Brazil, where they painted the Pedra Furada paintings. It was only when a wave of newcomers from Asia reached America that the Aboriginal relatives lost their status as absolute rulers of the continent. Over time, they were absorbed, assimilated. After millennia, only a genetic trace remains, confirming the Australian lineage of the first arrivals.

There are many indications, therefore, that Niede Guidon may be right to link the images of Pedro Furada with the Australian Aborigines. In this sense, they were the first to populate America 40,000 years ago.

A completely different clue is the latest hypothesis of Dr. Guidon and her successors studying Pedro Furada, which they shared in 2014 with journalists from the New York Times.

According to this theory, in the highlands of eastern Brazil there are even older traces of human presence, around 100,000 years old. This time, America's oldest explorers were to come by sea, not from Asia or Australia, but from Africa. But that's a topic for a completely different article.

Pyramid in the Bermuda Triangle

The Bermuda Triangle is famous for unexplained disasters and the disappearance of ships and planes, but this is not the only mystery of this reservoir. Well, one scientist participating in an oceanographic expedition claims to have found a pyramid standing on the ocean floor within the Triangle. And this is not the first time that structures resembling buildings have been found in the waters of that region.

Ancient messages and myths of many peoples from different parts of the world mention floods that destroyed ancient civilizations.

Only that in the area of the ocean floor of the Bermuda Triangle, scientists have not discovered - so far - the remains of human cultures, and no Indian mythical tradition tells about them. Will the new find change this?

At the end of March 2012, numerous news outlets reported the discovery of a mysterious pyramid-shaped object at the bottom of the Atlantic Ocean. This structure was to be identified by sonar by Dr. Verlag Meyer, an oceanographer who participated in the American-French expedition to explore the seabed in the Bermuda Triangle. Meyer claims that the pyramid is 2 kilometers deep, that it is 200 meters high, and that its base is 300 meters long. The explorers, using a remote-controlled camera, took pictures of this structure, which shows that its walls are perfectly smooth, and moreover, they have not even been overgrown with algae. Their surface reflects light so strongly that researchers had the impression that a glow was emanating from inside the building. Therefore, Meyer's colleagues concluded that the pyramid was made of a glass-like substance, which made it partially transparent.

Where did the pyramid come from at the bottom of the ocean, and it was covered with a glass lining? A working hypothesis was that the structure was originally located on land, but in the past there was a gigantic earthquake that permanently changed the topography. Water flooded the area where the structure was erected, so it is now at the bottom of the ocean. Is the hypothesis of submarine researchers likely to be confirmed geologically and archaeologically? After all, the alleged pyramid is located two kilometers below the surface of the water, so it is impossible to examine it more closely. Moreover, reports of this type are usually considered humbug. It turned out, however, that American scientists from nearby Florida got seriously interested in the matter.

"If we assume that the oceanographers' reports are true, then the pyramid would have to be built between 10,000 and 6,000 BC, because the continental shelf was not flooded at that time," said Dr. Calvin Jones of the Florida Department of Historical Resources.

Similarly optimistic about the discovery was Dr. Glen Doran, an anthropology professor at Florida State University.

- The earliest known to science pyramids date back to the 3rd millennium BC. He explained to the media. - So the structures discovered on the seabed off the coast of Florida would be at least four thousand years older than the known pyramids.

So if the mysterious structure was really erected by man, it must have been at least eight thousand years ago. And is there a rational explanation for the oceanographer's observation that it was made of glass? One would have to assume that it was covered with sheets of obsidian, or volcanic glaze, a material often used by pre-Columbian cultures. You can also consider slabs made of some kind of quartz.

When taking pictures, the object was illuminated with lamps, and you have to remember that light in water behaves differently than in air, and it refracted on smooth surfaces - and therefore an effective glow was created around the building. It is this picturesque effect that indicates that it may actually be a structure built by man, because science does not know any examples of creating such large smooth surfaces by the forces of nature.

The alleged pyramid discovered by Dr. Meyer is not the only structure at the bottom of the Atlantic, considered by some researchers to be the work of human hands. Apparently - because reports of this discovery have no verified basis - in 1977 another pyramid (more than 200 meters wide and high) was discovered in the region by an expedition led by Arl Marshall.

Equally uncertain sources indicate that the expedition of Tony Benik came across the next pyramid (as high as 330 meters) in the 1980s. The surface was also described by the explorers as "made of crystal" that reflected pale light.

Another - and equally controversial - find was reportedly found off the coast of the Bahamas, and therefore also in the Bermuda Triangle. In 1968 (according to other sources - in 1970) a certain Ray Brown, a professional healer from Arizona, discovered pyramid-like objects there. Brown claimed that when he was diving with friends in the Bahamian Bari Islands, he broke off briefly from the company. It was then that he noticed the mysterious structures. They were supposed to be at a relatively shallow depth, i.e. 20-30 meters. The Arizona healer kept his discovery a secret, but returned after a while with a team of four professional divers.

Only then did he examine the objects on the ocean floor. According to Brown, there was an entire cluster of buildings under the water, because apart from the pyramid, several domes and rectangular buildings connected by roads were identified. The divers also found loose artifacts - unidentified metal instruments and a statue of a man holding a large crystal in his hands, containing miniature pyramids.

Reporting his discovery, the expedition leader assured that the walls of the pyramid were smooth as glass and resembled a mirror.

Brown also found an entrance to the object so he could examine it from the inside. Inside there was a narrow corridor leading to a small rectangular room with a pyramidal vault, thus reflecting the shape of a pyramid. The walls of the submerged chamber were not covered with algae or other sea plants. Although the amateur diver did not carry a flashlight, he could see everything well because the room was lit, although Brown could not determine the source of the light. In the center was a metal rod to which a huge red crystal was attached. The basis of this structure was a carved stone slab on which was placed a pair of human hands made of metal, holding a crystal ball the size of a tennis ball.

According to the healer, the metal hands were blackened and burned, as if they had been exposed to high temperatures.

- I wanted to take a rod with a red crystal, but I did not manage to separate it from the ground - this is how Brown related his adventure in one of the interviews. But I easily took the crystal ball out of my metal hands. I left the pyramid with this object.

Ray Brown claimed that his expedition also retrieved metal devices of unknown purpose from the sunken city. All of these artifacts were donated to scientists at Florida State University. Apparently, the experts only managed to establish that the crystal strengthened light and electrical energy. The Arizona Explorer did not disclose his discovery to the world until 1975. Importantly, he never returned to a sunken city in the Bahamas. Why?

"When I was coming out of the underwater pyramid, I felt the presence of someone invisible," explained Brown. - I also heard a voice ordering never to go back to this place.

What about mysterious metal artifacts reportedly donated to university scientists? Apart from the discoverer himself, no one has confirmed that these items existed. Even Dr. Glenn Doran from the aforementioned university, who spoke officially and most seriously about the pyramid discovered much later by Dr. Meyer. Brown maintained that he had kept only the crystal ball to himself, which he had repeatedly shown to the public. In the photos, this item looks like an ordinary glass ball that you can buy in esoteric or souvenir shops. Have scientists studied it? Probably not, as a rule they avoided commenting on this alleged find.

In turn, some researchers of mysterious phenomena suggest that no one wants to comment on the monument, because Brown has been the victim of a global conspiracy to hide the truth about our species' past.

Conspiracy theories aside and attempts to analyze the esoteric properties of the crystal ball aside, a fundamental question must be asked. Assuming that the Meyer, Marshall, Benik, and even Brown pyramids are actually existing objects (and not of natural origin), who built them?

Neither in the USA nor in Central America have been discovered - at least so far - traces of a technologically advanced culture existing as many as eight thousand years ago. The oldest New World pyramids are about 4,600 years old, and they were built in the ancient urban center of Caral (Norte Chico culture) on the coast of central Peru, several thousand kilometers southwest of the Bermuda Triangle. The alleged underwater structures are even more distant from the Egyptian pyramids, which are as old as their "relatives" from Caral. It is therefore unlikely that between any of these two centers and the Bahamas existed in such a distant antiquity cultural contacts that allowed the idea of pyramids to travel from one end of the world to the other.

Moreover, it should be noted that even if we accepted the authenticity of the Triangle structures, they would have to be built much earlier than both the birth of Egyptian civilization and the oldest Peruvian culture...

However, assuming that the pyramids of the Bermuda Triangle were indeed created at least eight thousand years ago, why has the memory of their creators been lost? Has the waters of the Atlantic actually wiped out most of the traces of the enigmatic culture, and the few remains of it are on the ocean floor? Theoretically, it is possible, because the history of scientific research indicates that many cultures have disappeared from the face of the earth as a result of natural disasters. An example is the Indus Valley civilization, the traces of which were discovered in the nineteenth century, and the causes of its destruction were only recently established.

It was founded more than five thousand years ago, and in the heyday (from the middle of the third to the middle of the second millennium BCE), it probably formed one large state, which covered an area of up to 1.5 million square kilometers - from western India, throughout Pakistan to eastern Afghanistan.

There are ruins of many cities, including several huge ones, the most famous of which are Harappa and Mohenjo-Daro. In them, canalized, two-story houses, strongly fortified fortresses and large port quays were erected. After the Indus Valley culture, great works of art, elaborate jewelry, and even writing that have not yet been read have survived. So we know a lot about this civilization, but almost 200 years had to pass since the discovery of its first monuments in the 1820s for scientists to be able to determine how this great culture collapsed.

This mystery was clarified only in May 2012 by an international team of scientists. Researchers analyzed satellite images of the area above the Indus and soil samples there, which revealed changes in the relief of today's Pakistan over the past several thousand years. It turned out that around 1300 BCE a gigantic flood took place there, which permanently changed the topography. Metropolises, villages and farmland have been destroyed by water. The Indus Valley civilization had no chance of being revived, because in a short time its area suffered from global climate change that triggered a series of ecological disasters.

Over the course of a century, most of the rivers dried up and the fields turned to deserts. Such climatic conditions remained there until our times, and the sands guarded the monuments so well that the existence of a great culture was remembered only in Hindu mythical epics, which for a long time no one took seriously as historical sources.

There are many examples of cultures that have been destroyed by natural forces. Most of them were also forgotten by the people living in those areas, and discovering lost civilizations has only become the work of modern archaeologists. Thus, the Indus Valley culture is not an exception, which means that reports of unusual finds, even from places as remote from ancient centers of civilization as the Bahamas, cannot be dismissed in advance.

In this context, is the hypothesis that eight thousand years ago a similar fate befell the culture of today's continental shelf off the coast of America is just a fantasy?

After all, it is known that it was in that era - that is, between the 10th and 7th millennium BC - there have been a series of cataclysms on a global scale. All of them were associated with the end of the Ice Age and considerable tectonic movements and volcanic eruptions. The water level of the world ocean has risen so much that the waters have separated the British Isles from the mainland, the land in the North Sea (so-called Doggerland by scientists) has been flooded, and the Black Sea and the Mediterranean Sea have significantly increased their area.

Knowing the effects of this global deluge, which gradually increased over the millennia, it can also be assumed that it is responsible for the disappearance of the pyramid recently discovered by Meyer at a depth of two kilometers. If this structure was erected in a valley 10 or 9 thousand years ago, then when the ocean was flooded the valley, the pyramid could indeed have reached such a great depth. Is the structure discovered by Dr. Meyer a remnant of a civilization whose monuments have been swallowed up by the Atlantic?

Or maybe the "pyramid" is a natural creation? No answer to these questions yet.

Sunken Sunda Pyramid

The place which I am going to tell you about was considered sacred by the local people "always". Archaeologists got there during routine field research. In the area of the village of Karyamukti, miracles were not expected, only finds typical of those areas. However, the ruins of an extremely old pyramid were found, and when information about it was made public, the world also learned about the existence of an ancient culture and a sunken Sunda subcontinent. Because everyone has heard of Atlantis, but who apart from geologists knew about the existence in the distant past of Sunda, a huge peninsula in Southeast Asia?

Let's start with the day the unusual find was revealed. On November 5, 2012, the Indonesian daily Jakarta Post reported the discovery of a complex of ancient structures at the Gunung Padang archaeological site. This place is located at the foot of Mount Padang near the village of Karyamukti, which is located in the Cianjur district in the Indonesian province of West Java. On the gentle slope of the mountain, in February 2012, the ruins of an extensive stone structure were found, hidden at a depth of 3 to 12 meters. Numerous monuments were also found that did not have the characteristics of artifacts known from other cultures of this region. These are mainly machined large stones with symbols engraved in the shape of tiger claws or daggers. Whether these stelae decorated the large building or had a different purpose - for example cult - is still unknown.

The discovery surprised archaeologists so much that they kept it a secret for over half a year before confirming the age of the remarkable structure.

The research was conducted in two research centers - Indonesian and then American. Analyzes of soil samples, charcoal and various finds have shown that the Javanese building was erected in the 14th millennium BC.

"The results obtained by the Beta Analytical Laboratory in Miami dispel any doubts about earlier dates by the Indonesian National Nuclear Agency," said Dr. Budianto Ontowirjo, a geologist on the research team, in a statement to the Jakarta Post. - The structure on the side of Mount Padang is much older than the Giza Pyramids.

The issue of the Gunung Padang ruins was taken so seriously that in the fall of 2012, Susilo Bambang Yudhoyono, President of Indonesia took patronage over the research. On the part of the head of state, this step was not only a gesture to support science, but also an opportunity for Yudhoyono to gain political capital before the upcoming presidential elections. And the politician could gain a lot from taking care of the excavations in Gunung Padang, because these studies became famous all over the world overnight. Why? Previously, science assumed that no significant ancient civilization arose in the Indonesian-Malaysian region.

Meanwhile, the new discovery not only proved that large stone structures were erected in those areas, but it was done much earlier than anywhere else on Earth.

On behalf of the president, Andi Arief, Yudhoyono's advisor on ... disaster relief and social assistance, took care of the research project. Arief's function and education had nothing to do with archeology, but he was chosen not by accident, as he was known for his good contacts with the media and his image building skills. Arief treated the discovery as a prestigious matter and, in his interviews, emphasized that the Gunung Padang building was proof of his country's ancient roots and that Indonesia was already inhabited by a people culturally superior to other peoples in Asia and even the whole world in the distant past.

The politician's pompous declarations provoked the world of science to react. Many experts, including Indonesian, have questioned the idea that this position is the work of human hands - according to them, "ruins" are a cluster of natural rock formations.

In turn, a group of other researchers raised doubts as to whether the megalithic pyramid - as its discoverers called the Gunung Padang structure - was fairly exposed. However, contrary to these doubts, the discoveries made in the following years on the slopes of Mount Padang (as well as the verification dating of previous finds) showed that the building is really very old. Even older than originally assumed.

Initially, it was not known what shape the Gunung Padang construction was. Since preliminary studies showed that it was erected with stones, wood and earth, it was assumed that it was something like a giant burial mound. The results of subsequent research, published in January 2014, significantly changed this interpretation. GPR radars were used to collect the data, and these showed that the building was erected in the form of a pyramid, and only as a result of thousands of years of weathering, the shape of the building was eroded, which made it resemble an irregular mound. Images obtained from GPRs also helped locate many chambers and shafts hidden deep under the pyramid, the remains of which are now perfectly visible in the form of the remains of Cyclopean stone structures. However, this is not the end of the secrets discovered in this place.

During the subsequent excavation seasons, new dating was also carried out, which allowed to shift the construction of the building into an even more distant past.

According to information published in October 2014, the pyramid was last repaired and partially expanded about 13,000 years ago. However, the oldest pyramid - later constituting the core of successive stepped structures built up on it - was erected 23,000 years ago.

These dates, bewildering from the point of view of the history of human civilization, make Gunung Padang the oldest large structure in the world (at least from the point of view of present knowledge). It should also be considered a structure that has been used the longest, for the entire 10 millennia. Until now, no one had suspected that in such a distant past there was a society capable of designing and building such an object.

Not only the age and size of the Indonesian pyramid can amaze you. The technology used to erect this structure is also amazing. The research revealed the surprising expertise of the builders Gunung Padang. Many of the fragments of this titanic structure unearthed by archaeologists have been reinforced with some kind of cement. The binder used in the pyramid as a mortar consisted of 45% iron ore, 41% silica and 14% clay. Such a binder ensured a very strong bonding of various types of materials, and today it is a surprising proof of the high level of advancement of the building technique of the Sunda culture.

I mentioned Sun at the beginning of the article. The pyramid in Gunung Padang is located in the area where millennia ago stretched the subcontinent that does not exist today, known as the Sunda or Sundaland by scientists.

Unlike Atlantis from the legend of Plato or Lemuria and Mu, described by esotericists, quite a lot is known about the once real Sun only that this knowledge is limited to specialists in several fields, mainly geologists, botanists and zoologists. When discussing the pyramid in Gunung Padang, it is worth presenting some scientific findings about this sunken subcontinent.

Sunda was a huge peninsula that, between 100 and 13,000 years ago, included today's Indochina and Malay peninsulas, the islands of Borneo, Sumatra and Java, and areas that are now flooded by the South China and Java seas.

It was a subcontinent with an area comparable to Australia today. In that era, the world ocean was much lower than today, even by about 130 meters. However, with the last stage of the Ice Age, the water level rose gradually, flooding the lowest parts of the land. On the other hand, the jump (from 50 to 70 meters) of water increase took place in just one millennium, that is 13-12 thousand years ago, when the shape of the land known to us today was formed. According to the comparison of the dates obtained at Gunung Padang with the dating of geological changes, shortly after the last time people rebuilt the Javanese pyramid, the last lowlands of the Sundas were flooded, and seas and straits were formed in their place. Only the former highlands and mountains, which are now islands of Indonesia and Malaysia, protruded above the waves. A huge archipelago was built on the site of Sunda.

The catastrophe not only changed the geography of this region of the world, it also buried ancient cultures that flourished in this lost subcontinent. Until recently, it was not even suspected that any ancient advanced culture existed in this part of the world. Moreover, it was not expected that anywhere in the world monumental buildings were erected so long ago. The reason was simple: there was no evidence of their existence in such a distant era.

In addition, the area of the former Sunda is a very difficult area for archaeologists. First, much of the subcontinent has been flooded and underwater research is extremely difficult, risky and costly. Secondly, even with these monuments that were located in the areas of today's islands, time and climate have been ruthlessly handled - with high temperatures and air humidity prevailing there, organic material deteriorates rapidly, and even stone erodes very quickly. The remains of humans and animals, products made of bone, wood and leather are degraded by microorganisms at a rapid pace. Let's add to this the action of the green element - the jungle. The buildings are overgrown with vegetation, roots and creepers are even bursting with large boulders used to build the structure. The wonderful monuments of ancient Egypt, Sumer, and even the much earlier monumental buildings of Göbekli Tepe, Karahan Tepe and Tell Qaramel have survived to this day only thanks to the dry climate of the Middle East and North Africa. The desert sand has perfectly preserved artifacts from millennia ago. However, completely different - and disastrous for monuments - conditions have prevailed in Southeast Asia for millennia, so only modest traces of the Sunda culture have survived.

Add to that the slim chance of success in the field of underwater archeology. It is unlikely that any ruins will be preserved on the bottom of the seas and straits of the Indonesian-Malaysian zone, which were once valleys of the submerged subcontinent, because this area is tectonically unstable.

Over the last several thousand years there have been many earthquakes and tsunamis. How destructive these forces are, was demonstrated by the tsunami caused by undersea quakes in 2004, which took place precisely in the area of the former Sunda. Over a quarter of a million people died then, many port towns were destroyed, and the Sumatra coastline shifted by 30 meters.

So when we take into account the passage of time, the effects of climate and vegetation as well as natural disasters, we understand that the ancient finds from the area of the former Sunda should be considered unique. Therefore, it is a real scientific miracle that, in addition to the pyramid from the hillside of Padang, archaeologists have also managed to find other works of the forgotten Sunda culture.

Interesting cave carvings and rock paintings were discovered in East Timor, a small island state neighboring with Indonesia (the area there also used to be islands situated just off the coast of Sunda for millennia). This find was also found by accident, because during the search for the bones of extinct giant rats. In early February 2011, the science service "Science Daily" reported that an international team of archaeologists and palaeontologists led by Dr. Ken Aplin from the Australian State University of Queensland in Brisbane revealed the discovery of unusual artifacts in the Lene Hara cave in northeast Timor. In May 2009, scholars wandered to one of the sides of the grotto, and there they looked into the eyes of the ancient Timorese.

"We focused our attention on the ground, looking for fossilized remains of large rodents, but when one of our colleagues directed the flashlight upwards, we saw the carvings of human faces facing us," Dr. Aplin said. "These images were convex like bas-reliefs, so we had no doubt that we discovered monuments of prehistoric art."

The Lene Hara petroglyphs show stylized human faces with well-defined details: eyes, nose and mouth. Some shows have red ocher elements. One of the bas-relief faces "wears" a round headdress with rays that frame the image. Analyzes of ocher samples from this particular artifact, carried out at the University of Brisbane laboratory, showed that the "sun-kissed face" was no less than 12,000 years old.

This dating made researchers realize that they have to look differently at finds previously made in the same place. Well, scientists first found their way to the Lene Hara cave in the early 1960s. There they discovered numerous rock paintings depicting animals, people, boats and various signs generally defined as decorative motifs.

In those days, however, it was not possible to date these pictures, it was only assumed that they were at most several thousand years old, i.e. they come from the era when the first arrivals from India arrived on the islands.

Only in 2009, on the occasion of the discovery of the "face gallery" by searchers of extinct rats bones, samples of the pigment were collected, which were used to make paintings known for several decades.

A year and a half later, Dr. Sue O'Connor of the Australian National University tested the small amounts of red ocher. The obtained result surprised the world of science, as it turned out that the ocher layer was placed on the limestone wall of the cave over 30 thousand years ago!

Similar rock images have previously been discovered elsewhere in the island region, but their age has not been determined by physicochemical analyzes. It was stubbornly assumed that they were made by people who had settled on the islands from India during the last few millennia.

It was only after the findings of the Australians that it was suspected that these artifacts might be as old as the paintings from Lene Hara.

- Dating Timor's rock art should be a priority in future research due to the cultural significance of these finds - said Dr. O'Connor in one of the 2011 press interviews. - Only now are we beginning to understand the distant past that art developed in this area.

Also in 2011, Sue O'Connor established the age of another ancient find from East Timor. It was a fish hook made of shells discovered in the Jerimalai Cave. It has proven to be the world's oldest human-used fishing hook. It was made between 23 and 16 thousand years ago, and therefore also in the heyday of the culture next to Sunda.

Let's go back to the pyramid in Gunung Padang, one of the oldest (over 20 millennia), and certainly the largest artifact from the lands of the lost subcontinent. For comparison, cave paintings from Africa, France and Spain have long been considered the oldest art monuments in the world. There were often problems with the dating of these works by ancient artists, so it was carefully assumed that the oldest paintings - probably more than 30,000 years old - were found in northern Spanish caves such as Cueva de Altamira and Cueva de El Castillo. The art of Asia, the Indian subcontinent and the Indonesian islands seemed to be extremely young compared to the African and European art. Consequently, no one suspected that in the past as distant as the time of the creation of the Cueva de El Castillo works, there were also nameless artists from the peoples inhabiting the today sunken subcontinent. Another discovery from the region near the Sunda changed this state of affairs.

In early October 2014, a team of researchers from the Australian Griffith University in Queensland, led by Dr. Maxime Aubert and Dr. Adam Brumm, published the results of research on cave art from the islands of Indonesia. Thanks to a series of analyzes, it was found out when these paintings were created, and these findings surprised the world of science. It turned out that some of the paintings are not less than 39,900 years old! Australian scientists looked closely at the rock art works from a cave near the town of Maros on the island of Celebes, which lay just off the coast of Sunda for millennia.

The dating of subsequent works was also successful, because the experts were able to use the method of measuring the layers of calcite dripstone on the paintings formed by limestone deposited with water.

Twelve rock representations of human hands, which prehistoric artists made by blowing paint through a bone tube around a hand pressed against a cave wall, were tested. Two images of animals were also analyzed. It turned out that these works are the oldest paintings in the grotto (they are about 40 millennia old), while the others are much younger, but still very ancient, because they were made 27,000 years ago.

The findings of scientists from Griffith University mean that the ancient people of Celebes created paintings in one place for several millennia. It is reminiscent of the activity of cave artists from Europe, because they, too, left their artistic traces in rock sanctuaries throughout the millennia. In Celebes, the inhabitants of which were probably in constant contact with the inhabitants of Sunda, an art developed comparable to its European peer, which left cave paintings from southern France and northern Spain. Thus, we can see that the cultures of the entire region of Sunda and near the Sunda took tens of millennia. It can also be carefully assumed that the central Sunda culture in its heyday was the most advanced of the world's cultures at that time, since it erected such a special and unique structure as the pyramid at Gunung Padang.

We know quite a lot about the origins and development of rock artists in Europe. Even DNA analyzes of their remains were carried out, allowing the appearance of the authors of these paintings after tens of millennia. But what peoples lived at the same time in the Sun and the surrounding islands, the generally understood Sunda region? Who created the artworks there and built the titanic structure in Gunung Padang?

Maybe the research on a certain tribe living in the middle of the former Sundy, considered to be a living fossil of humanity, will help to solve this puzzle?

In 2005, an international group of scientists led by Dr. Vincent Macaulay from the University of Glasgow discovered that in the hard-to-reach mountains of the Malay Peninsula (Malaysia) lives a strain with extremely archaic anthropological features. The DNA of the members of this tribe (which is an extremely isolated part of the Orang Asli peoples) resembles the genotype of African members of the Homo sapiens species from around 80,000–70,000 years ago, ie from the period of the extremely important migration from Africa to Asia and Europe. And this migration is so important because it is believed that the majority of the world's population comes from the individuals participating in it. This would mean that the representatives of the aforementioned mountain tribe belonging to the Orang Asli group kept in a "clean state" the genotype of migrants from Africa, and therefore are their only living direct descendants.

Also their appearance is typically negroid (black skin, wide noses, short, curly hair, short stature), which makes them stand out even among other Orang Asli tribes. Interestingly, the collective name of these tribes means "First People" in Malay, so they were also the ones who lived here "forever" for their neighbors. Unfortunately, although the mountain Orang Asli come from times so ancient that they could have been the creators of the Sunda paintings and buildings, no stories about this have survived in their myths or messages. This tribe has lost its original tradition, and their today's beliefs and traditions are a patchwork of cultural influences of neighboring - and much younger - peoples. Since it cannot be said whether it was the ancestors of Orang Asli who painted the pictures from the cave near Maros and built the pyramid at Gunung Padang, do we have other candidates? Or maybe those of the Sunda and surrounding area who created the high culture were not our species at all? Maybe they came from a different branch of the hominid family tree?

Two sensational discoveries made in recent years in the former subcontinent provoke such a hypothesis.

First, at the turn of 2003 and 2004, in the Liang Bua cave on the Indonesian island of Flores (similarly to Celebes or Timor, near the former Sunda), the remains of individuals from a previously unknown species, called Homo floresiensis, were discovered in a colloquial way today. . The found skeletons of these hominids and the stone tools produced by them were expended for a long period, from 90 to 12.5 thousand years ago. Analyzes have shown that the Flores man is most likely a stunted descendant of Homo erectus, a species of hominids that had traveled from Africa to Asia hundreds of thousands of years earlier than Homo sapiens sapiens.

The most characteristic were the size of the hobbits, as they was at most one meter tall and with body proportions different than modern humans. They lived 12,000 years ago, that is, in the era of the final destruction of the Sundas, although - as some ethnographers believe based on folk tales from Flores - the remnants of the Homo floresiensis community have survived to modern times.

So was this species dominant in the lost subcontinent and surrounding islands? It would be an effective culmination of discoveries about a lost culture.

Unfortunately, the fossilized hobbit bones were found only outside the reconstructed area of Sunda - in Flores, and also on the island of Luzon in the Philippines (which I will talk about a bit later).

While Flores was relatively close to the shores of Sunda, Luzon were really far away. However, the situation of the Hobbits as a "Sunda species" can be saved by research conducted for years in Sumatra, where scientists - following folk tales and residual material traces - are looking for a creature called orang pendek ("low man") by the natives. Anthropologists speculate that he may be identical to the human of Flores, and that would mean that the Hobbits inhabited not only the islands east of Sunda, but also the subcontinent itself. For now, however, there are no specific details on the extent of orang pendek occurrence, so let's return to the announced discovery in the Philippines.

In June 2010, a team led by Dr. Armand Mijares of the University of the Philippines in Diliman discovered 67 millennia-old remains of a hominid that physically resembled Homo floresiensis in Callao Cave on Luzon Island.

It must be emphasized that Luzon was an island also in the Sunda era, so it was only possible to get there by sea. The extinct dwarf hominids must have been skillful sailors, since in such distant times they sailed the long route from Flores to Luzon (or was it the other way round?) On boats or raft.

So, as we can see, the question of what kind of people formed the Sunda culture remains unanswered. Were they the ancestors of the mountain Orang Asli, or were they hobbits? Could the beings from Flores, measuring one meter in height and weighing up to 30 kilograms, be able to build structures as large as the pyramid at Gunung Padang? And would the primitive Orang Asli or the Homo erectus hobbits be able to create such works of art as the rock paintings of Celebes? Why not, archeology and anthropology surprise us with new discoveries every year.

Since when do people care for the health of their fellow tribesmen? How long have they been able to count time and build structures that help to observe the apparent motion of celestial bodies? How many thousands of years ago did they start building temples to worship deities in them? Is the realm of beliefs and symbols appropriate only to the human species? Answers are still being sought to these and many other questions concerning the history of spirituality, religion and social relations. Maybe we'll get closer to them, getting acquainted with the interpretations of a few puzzling artifacts that do not match the current knowledge about human development.

Caveman's tooth filling

In 1988, a grave was discovered in the Dolomite Mountains in northern Italy, located in the Ripari rock shelter under the Villabruna rock. The grave contained a well-preserved skeleton of a man living at the end of the Paleolithic and at the same time in the last stage of the Ice Age. As established during the analysis of the remains, it was a young man who died at the age of about 25. Scientists from the University of Ferrara took care of the discovery. In 1990, carbon-14 dating there showed that the bones were 14,160 to 13,820 years old. Only a quarter of a century later it turned out that the skeleton of the young man from Villabruna hides an extraordinary secret.

New electron microscopy research has revealed the advanced dental skills of our distant ancestors.

In July 2015, the journal Scientific Reports announced a discovery made by a research team led by Dr. Stefano Benazzi, a paleoanthropologist at the University of Bologna.

It was established that one of the young man's teeth was treated with a flint tool to remove caries. The tooth was drilled and ground - probably with success, because, as shown by the further wear of the tooth, it served this man for a long time, so the "dental" treatment was carried out long before the patient's death.

- The Villabruna discovery is the oldest example of a medical intervention to eliminate a disease condition. It is significantly ahead of the discovery of traces of dental surgery, represented by drilled teeth from the Mesolithic and Neolithic periods, around 9,000–7,000 years ago, explained Dr. Benazzi in a statement for ABC. - In previous studies, no traces of these treatments were found in the teeth of the man from Villabruna. Although the holes themselves were located, they were described as typical caries-induced lesions.

Benazzi did not mention the discoveries relating to the use of certain basic hygienic procedures, traces of which have been preserved from even earlier times, because of the early Paleolithic. These cases, however, involved the mere use of wood and bone toothpicks, traces of which were found on the teeth of Neanderthals 100,000 years ago.

On the other hand, abrasion or even drilling of teeth with specially prepared flint tools - like in the case of the Dolomite highlander - is a completely different level of medical activities.

As can be seen from the example from Villabruna, already 14 millennia ago, prehistoric "dentists" not only pulled out teeth, but also treated them. They had skills that European specialists only learned again 150 years ago. Toothache is a nasty feeling, what's more, tooth decay affects the entire body. It causes irritating inflammation and headaches. People of the Stone Age already suffered from it, although it must be emphasized that Paleolithic hunters, such as the man from the Dolomites, who ate mainly meat, suffered from decay much less often than people from later eras, who mainly fed on agricultural products.

Scientists assume that caries treatment was a multi-stage process even in ancient times like the Palaeolithic, as evidenced by the find from Villabruna. First, the diseased area was drilled, then it was cleaned of infected tissue, then an antiseptic filler, presumably herbal, was inserted into the hole, and finally the hole was covered with some substance that served as a filling.

How do you know, however, that in the case of a man from Villabruna, it is about the effects of dental procedures, and not about ordinary tooth losses resulting from lesions? It was demonstrated by the analyses carried out with the scanning electron microscope.

In this way, Benazzi's team concluded that inside the tooth (and thus in the drilling), specific traces of interaction with very small flint tools, the so-called microliths, were preserved. However, the experts did not stop at these conclusions. They conducted experiments as part of experimental archeology and, using tools from several millennia ago, drilled a tooth of modern man. Traces of the use of Palaeolithic tools turned out to be identical to those inside the teeth of a Dolomite hunter. On this occasion, scientists found out that dental treatments from 14 millennia ago were very advanced. They found that the Ice Age dentist used needle-diameter flint drills, as well as thin tweezers (presumably made of horn) and small tools for grinding and polishing teeth. And such complex instruments and advanced methods show that medical and surgical knowledge must have evolved for a long time, so its origins seem very distant and so far impossible to establish.

"The specialized tools used to treat a man from Villabruna show that the Paleolithic people already knew about the detrimental effects of caries on overall health and the need for invasive treatment to remove infected tissue," said Benazzi. - It should be emphasized that it would be much easier to extract such a tooth, because the treatment of a sick tooth is complicated and requires skills that modern dentists acquire during difficult studies.

Until the beginning of the 21st century, the oldest finding from Denmark was considered to be the oldest traces of teeth drilling for medical purposes. However, it turned out that dental skills were developed even earlier, as evidenced by a discovery in the town of Mehrgarh in Balochistan, Pakistan.

Many years of excavations were carried out first by the French archaeological mission of Jean-François and Catherine Jarrige, and since the last years of the 20th century by an American-French-Italian expedition led by Dr. Andrea Cucine from the American University of Missouri in Columbia. In Mehrgarh, the ruins of a great Neolithic settlement were found, which was founded 9,000 years ago, and during the period of the greatest development of the Indus civilization (4th-3rd millennium BC), it grew so much that it was one of the largest cities of this culture. People living in Mehrgarh grew grain and raised cattle, but the finds that attract attention are the unusual sculptures and jewelry made of shell, mother-of-pearl, amethysts and turquoise. Mehrgarh as early as the 9th and 8th millennium BCE they used specialized small tools, thanks to which the processing of raw materials, including semi-precious stones, was at a level that was achieved only a few thousand years later in ancient Egypt.

However, the real sensation was the discoveries about the dental condition of people from this culture, which was reported in the scientific journal "New Scientist" in 2006.

Archaeologists found human remains from 9 millennia ago, with traces of tooth treatment, which were precisely "drilled". It is possible that the treatments used were the same jewelry tools that were used to make the aforementioned jewelry. Experts from the University of Missouri determined that the drilled teeth did not contain a trace of caries, so the treatments brought the desired results. It is not known, however, what substance was used for the fillings, because it has decomposed over thousands of years. Dr. Cucine's team tentatively assumed that the holes in the healed teeth were filled by dentists from Mehrgarh with a natural bactericidal element, possibly of plant origin.

During the procedures themselves, the patients were probably administered painkillers, for example extract from the poppy growing in this region. According to earlier findings of archaeologists, simple hand drills, so-called drill bits, have been used for work in wood, stone or bone since the 4th millennium BC, but finds in Pakistan have shown that the idea of such a device - both for drilling teeth and creating delicate jewelry - was born much earlier.

What's more, Mehrgarh invented devices that are much more delicate and of better quality than the drills known to science for creating stone or bone tools.

This is evidenced by the dimensions of the holes drilled in the teeth - they have a diameter of only 1.3 to 3.2 millimeters and a depth of up to 3.5 millimeters. Dental treatments proved effective because the enamel around the holes was smoothed, which proves that patients ate hard food long after the treatment. Archaeologists have determined that nine millennia-old Mehrgarh dentists embedded small flint drill heads into long, thin pieces of wood - much like arrowheads were mounted on spars. It is assumed, moreover, that the drilling technology (whether in stones with drill bits or delicate drill bits in the teeth) was developed on the basis of the ability to construct and use a bow. This is indicated, inter alia, by finds from Pakistan.

The bow has been in use for at least 35,000 years, and since many small stone blades resembling arrowheads have been discovered in Africa, dating back 70 millennia, perhaps the invention of the bow occurred in such a distant past.

What is important for us here, the bow was the first device in history that stored mechanical energy (while drawing) and then returned it (while releasing the string).

It is not known when it was discovered that the arc could be used as a drive for a drill, but it was probably accidental. Such a drill was created when the string of the bow was wrapped several times on a drill with a flint blade (perhaps it was originally an arrow), and then the ends of the string were reattached at both ends of the bow. The man held the blade with one hand to prevent it from slipping while making the hole, and with the other hand he moved the horizontally held bow. This is how an efficient drill was created.

As it was established during the analysis of the finds in Mehrgarh, local dentists and jewelers used small arcs to drive the drills, which allowed them to achieve a speed of 20 revolutions of the drill per second.

Soon after Dr. Cucine presented the results of his research with Mehrgarh, it turned out that the ability to drill holes - and small, and therefore useful in dentistry - it is much older than the findings from Italy and Pakistan have already discussed. Interestingly, the discovery that I am about to tell you about, not only comes from a more distant era, but also relates to a human species other than ours.

The first remains of pre-humans called Denisovans were discovered in 2008, but the discovery was officially announced in 2010. So far, material evidence of the existence of this species comes from only one place - the Denisov Cave in the Altai Mountains, which is located about 170 kilometers south of the city Barnaul (Russia, southern Siberia, Altai Krai).

Two teeth and one finger bone of a hominid of a previously unknown species were found in the cave. The genetic material was obtained from these remains, and the first analyzes showed that the bones are over 30,000 years old. Subsequently, Siberian scientists invited the British expert in the field of geochronology, Dr. Tom Higham from the University of Oxford, who carried out further analyzes using the latest laboratory techniques. As reported on September 16, 2015, the site "Ancient Origins", it turned out that thanks to the method of analyzing the sequence of the genome, it can be assumed that the Denisovians arose as a separate species about 170 millennia ago, and disappeared 50,000 years ago. It is worth mentioning that other results came from the genome sequence studies carried out on the basis of the Denisov Cave bones by Dr. Svente Paabo from the German Max Planck Institute in Leipzig. He determined that the Denisovans and Neanderthals separated themselves from an unspecified African hominid line about 400,000 years ago, then formed one species for a hundred thousand years, and then the Denisovans separated from the Neanderthals.

As you can see, even the genetic research trusted so much today can lead to different conclusions, and we do not know who to fully believe, because both of these research centers enjoy great authority.

However, no matter how different the findings of the researchers, the discovery of a hitherto unknown human species is of course very important from the point of view of the development of our knowledge about prehistory. Here, however, I mean another issue related to the topic of the article.

In 2008, during excavations in the Denisov Cave, near the Homo denisovaensis bones lying in the geological layer dating back to about 50 millennia ago, stone tools and a stone bracelet were found. This last find is sensational, because in the ornament made of polished green chlorite from deposits located 250 kilometers from Denisov's Cave, an ancient "jeweler" bored a hole 8 millimeters in diameter.

The discovery surprised scientists as much as the discovery of a hitherto unknown species of humans. Information about the artifact and the research carried out on it was reported on May 7, 2015 by the British Daily Mail.

The delicate, partially damaged C-shaped bracelet was examined by a team of experts led by Dr. Mikhail Shunkov, deputy director of the Institute of Archeology and Ethnography of Cultures of Siberia of the Russian Academy of Sciences in Moscow, and Dr. Irina Salnikova, director of the Museum of Folk History and Culture of Siberia and the Far East in Novosibirsk, where the ornament is currently stored. The analyses showed unequivocally that the hole in the stone bracelet had been drilled - so a drill and drill had to be used. So we are talking about a discovery that radically changes our knowledge of the development of technology in the Paleolithic. After all, previously the oldest holes drilled in the teeth were those discovered in the remains of a young man who was buried in the Italian Dolomites 14,000 years ago.

"The find proves that Denisovans were more technologically advanced than Neanderthals and our kind who lived in the same era," Salnikowa told the Daily Mail. - In fact, nowhere in the world in such an early era was anyone using such advanced technology. This is a great mystery, because from the genetic and morphological point of view, Denisovans are much more primitive than Homo sapiens sapiens.

However, as we have already noted, research and experiments in the field of dental treatment had to be carried out much earlier than the first traces of teeth drilling at Mehrgarh and Villabrun. Otherwise, the "dentists" in these two localities could not have attained such medical proficiency or specialization in the manufacture and selection of instruments. However, science cannot indicate where and when the first "dentist" saw the first patient. So does the palm of precedence belong to the Denisovans? It is true that we do not know anything about their skills in the field of medicine, including dentistry, but it was Homo denisovaensis who used the drill 50,000 years ago, i.e. 10,000 years before our species came from Africa to Europe.

What about the caveman seal mentioned at the beginning of the article? Unfortunately, although the Palaeolithic healer of Villabruna must have used one, no trace of her was found in the teeth of a Dolomite hunter. In turn, according to the assumptions of archaeologists studying the finds from Mehrgarh, the local dentists probably made fillings of bituminous mass or resin, because these substances can be quite durable fillings, and they work aseptically. By analogy, it is assumed that the seal of the Villabruna youth was made of some other natural substance, for example beeswax, the use of which is indicated by a similar, although much younger (6.5 thousand years old) find from the area of Slovenia. The remains of an old man were discovered there, in whose healed tooth there were remains of wax, which, in addition to its filling function, also has bactericidal properties.

Stone calendar on Mount Sunduki

In the central part of southern Siberia, in the territory of the Autonomous Republic of Khakassia (which is part of the Russian Federation), rises Mount Sunduki. The hilly steppe surrounding it is now almost uninhabited, but it was different 16,000 years ago. At that time, a community lived there, which built a large and clearly visible stone structure on the Sunduca summit. Russian archaeologists who found this object in early 2013 claim that it is the world's oldest simplified astronomical observatory.

It is eight deliberately placed rectangular sandstone blocks, forming - as the discoverers believe - a calendar.

If their interpretation is correct, it would be the oldest calendar known to science. Thanks to properly positioned boulders, each of the then inhabitants of the area could independently determine the time of the year and even the day, observing the position of the Sun in relation to the stones at the summit of Sunduki. Such knowledge was needed by the paleolithic people, for example, in order to find out when the months of the wanderings of herds of wild animals that were hunted would come.

But are Russian scientists right to guess that a huge stone artifact dating back 16 millennia is a calendar? After all, it is assumed that the first calendars were invented only after humans had been involved in agriculture for a long time - date setting methods were to be developed in the Middle East some 6,000 years ago to help determine the sowing or harvesting times of cereals. However, the discovery of a megalithic structure in Siberia suggests that perhaps ways to calculate the passage of time were invented much earlier and for other reasons.

The perception of time and its measurement is mainly influenced by three phenomena: the period of the Earth's rotation around its own axis, the period of the Earth's circuit around the Sun, and the apparent changes of the Moon's size.

The rotation of our planet lasts 24 hours and the circulation around the central star of the system takes about 365 days. In both cycles, there are constant elements that influence our life, as perceived by the man observing the position of the Sun in the sky: the division of the day into day and night and the division of the year into four seasons. Another point of reference is the phases of the Moon which determine the passage of months.

But when did time "start"? Subjectively for each of us at birth. For the universe, on the other hand, this moment occurred several billion years ago. However, billions of years are too huge a range of time for a human being to deal with during everyday calculations. Therefore, time measures were needed that were proportional to the length of an individual's life and to the history of the people from whom the creator of the calendar came from - as a result, many different calendars were invented, depending on the culture in which their creators functioned. In other words: each culture measures time in its own way. The calendar currently in force in Europe and the entire Western culture, the so-called Gregorian calendar, introduced in 1582 by Pope Gregory XIII, takes the approximate date of Jesus' birth as the "zero" point. The years relating to the events that occurred earlier, i.e. before our era, began to count only in the 18th century.

In turn, biblical assume that the world was created just over 6 thousand years ago, and it was then that time began to flow. In the Jewish calendar, reformed in the 11th century A.D., the years are counted from the date of creation of the world that is from October 7, 3761 B.C. According to the Islamic calendar, time is counted from July 16, 622 A.D., i.e. from the date of the new moon closest to the date of Mohammed's escape from Mecca to Medina.

However, these are all calendars drawn from a common source, and the beginning of time recorded in them is relatively recent. But who, however, and when did they define and, in a way, "discovered" the concept of time? Who and when felt the need to divide the river of time into days, months, and years? There are many indications that the calendar is an invention of the people living at the end of the Paleolithic period.

I have already mentioned that science for a long time linked the concept of inventing the calendar with the transition of humanity to advanced tillage in the Neolithic era. The views on the origins of the calendar began to change gradually only with the hypothesis of Dr. Michael Rappenglueck of the University of Munich, published in 2000. According to him, people used lunar calendars as early as the end of the Palaeolithic period, as exemplified by some of the younger, 16,000 years ago, rock paintings from the famous Lascaux Cave in France and the even younger (14,000 years ago) "map of the sky" from the Cueva de El Castillo Cave in Spain.

"This is a representation of the constellation called today the Crown of the North," Rappenglueck said. "In the Lascaux cave, on the other hand, the dots drawn by the paintings of the bull, antelope, and horse symbolize the 29-day lunar cycle."

For years, the German researcher was alone in his views, as science assumed that people began to observe the passage of time and create calendars much later, in Neolithic times, as evidenced by the Sumerian civilization monuments from Mesopotamia or the structures of various megalithic cultures, such as the famous Stonehenge stone circle. Most scientists believed that such buildings were the first in the history of mankind to serve as astronomical observatories (as well as and monumental calendars) and, at the same time, places where religious rituals were held.

Many megalithic objects related to observations of celestial bodies were discovered - from Europe and Africa to the ends of Asia - but they were similar in age to Stonehenge, or at most 4,000 years old. A new look at the oldest megaliths was provided by the discoveries from the turn of the 20th and 21st century when in the Nabta Playa region in southern Egypt the Polish-American mission encountered typical megalithic stone astronomical observatories in the form of circles and ranks of steles directed at the most important star systems. This huge megalithic complex began to emerge 7.5 thousand years ago, and it was the work of the existing in that area at least two thousand years earlier, a dynamically developing culture of Neolithic farmers (considered to be the direct predecessor of Egyptian civilization).

But even the Nabta Playa complex did not turn out to be the oldest discovery related to prehistoric star gazing, time measurement, and calendar creation.

The next significant discovery - the stone structure from the top of Mount Sunduki mentioned at the beginning of the article - showed that the passage of time was probably calculated even earlier: just in the period indicated by Dr. Rappenglueck. In April 2013, the journal "New Scientist" presented the results of the research of the Palaeolithic Observatory in the Khakassia part of Siberia. The analyses were carried out by the same scientists who discovered the monument shortly before, i.e. a team led by Professor Vitaly Larichev of the Institute of Archaeology and Ethnography of the Siberian Branch of the Russian Academy of Sciences.

The information services took an immediate interest in the matter, but at the same time simplified it very much, reporting on the discovery of "the world's oldest astronomical observatory" and "Siberian Stonehenge". During this time, scientists from many fields began to have a serious discussion about the importance of the Khakassic find. Some considered it to be a breakthrough leading to a change of views on the development of mankind at the end of the Paleolithic period, but others considered the interpretations of Russian archaeologists to be too speculative and too far-reaching. However, the discoverers of the Sunduki construction themselves had no doubts about the great importance of their find.

"The so-called Siberian Stonehenge is the oldest example of a place from which people consciously and with the use of tools observed the sky and counted the passage of time," professor Laritchev said.

Nowadays, it is possible to argue whether the calendars indicated by Dr. Rappengluck painted on the walls of the caves are a thousand years younger or older than the monument from Sunduki, but such details do not matter. It is important whether the age of the Siberian observatory and its role is confirmed by other archaeological findings made in this place. The distant time of the creation of the stone calendar is indeed indicated by the numerous objects discovered in the area characteristic of that era and the art of late Palaeolithic rock carvings depicting various animals, as well as people - perhaps builders of this unusual construction. However, the crowning evidence for the function of the building and the time of its erection are to be found in the analyses of experts in the field of astroarchaeology. Computer simulations of the view of the sky from 16 thousand years ago show that the system of boulders on the top of Sunduki could indeed enable proper observations of such a system of celestial bodies as it existed in that distant epoch.

"Ancient inhabitants of the valleys surrounding Sunduki could observe daily how the sunrises and sunsets and moon quadrants run in relation to their observatory," explained professor Laritchev. "When assessing this find, we must remember that the first known sundials, found by archaeologists in ancient Egypt, are just over 3.5 thousand years old."

The hypotheses of Laritchev and Rappenglueck indicate that the knowledge about the passage of time and its counting could be spread already at the end of the Paleolithic period, and was used by peoples living in a vast area from Spain and France to Siberia. Since then people have developed many methods and devices to measure years, days, hours. However, every day, man still subjectively feels the passage of time, so universal measures and tools are necessary for him, especially today, in the age of civilization based on technology, because subjectivity significantly affects the perception of time. Interestingly, its passage is closely related to the size of the observer's body - which in turn is influenced by the process of evolution. According to research results published in September 2013 by Irish scientists, among all animal species, the perception of time is closely related to the size of the individual.

The smaller the animal, the faster its metabolism, which leads to faster reactions. For this reason, it seems to the tiny creature that the behavior of animals of much larger sizes is slower. The team investigating this problem, led by Dr. Andrew Jackson from Trinity College Dublin, stated that this principle also applies to humans.

"The inspiration for this study was a daily observation," Jackson explained. "I was probably not the only one who had the impression that small children seem to be doing everything in a great hurry. It turns out that it is not a matter of impressions, but of noting a fact. For toddlers, time goes by differently than for adults. The subjective assessment of the passage of time changes with age. It reaches a stable level only when the human body becomes mature, so it reaches a certain order of magnitude, appropriate for our species."

Irish scientists have analyzed the behavior of more than 30 species, not only humans, but also rodents, lizards, eels, hens, pigeons, dogs, cats, turtles, and flies. It turned out that the smaller and more agile the creature, the greater its ability to perceive is in a certain unit of time.

"The signals received visually by small organisms are processed by the nervous system faster than in large organisms," said Dr. Graeme Ruxton of the University of St. Andrews in Scotland, commenting on the results of this research. "The brain capacity of even the smallest animals is really impressive."

The fly is certainly not a thinker, but it makes good decisions for itself much faster than an extremely intelligent person can. So time flows completely differently for a man measuring 180 centimeters and weighing 80 kilograms than for a three-millimeter fly weighing milligrams. According to scientists, a fly sees events in the world around it four times faster than a human being - and that's a good thing. According to scientists, such large differences in perception of the passage of time are the result of evolution. Small creatures (permanently endangered by the much larger ones) have developed a remarkable survival strategy over billions of years - to avoid being eaten, they live "at a different time" from their potential killers, or more precisely: they exist at an accelerated rate. For small animals, it is a matter of life and death to perceive time not according to a rhythm measured by clocks and calendars, but according to a subjective feeling.

Man has never encountered a predator that he would not be able to defeat with a weapon he invented, so (not counting childhood age) he relies only slightly on a subjective evaluation of time. However, it can be assumed that people - developing culture, taking advantage of the world of animals and plants, inventing various useful technologies, and consequently creating increasingly complex communities - were doomed to develop methods for measuring time.

New discoveries and a fresh look at old findings suggest that time measurement has become important for man as early as several thousand years ago, so the calendar is not an invention of Neolithic farmers. It may be suspected that time measurement methods have already been developed by the Paleolithic hunters wandering behind herds of wild animals. The development of specific time units could also be useful for the first Stone Age sailors, wandering around the Mediterranean on boats.

Was it then, in such a period of prehistoric history that people discovered what time is, and then sanctified it? The giant artifact at the top of Sunduki shows that already 16 thousand years ago man was ready to devote a lot of energy and time to create sanctuaries for time.

Adam's Calendar

"The so-called Adam's Calendar may be the oldest megalithic building in the world," says Michael Tellinger, a writer and researcher of the past of his country, South Africa. An unusual stone circle with wavy inner walls, called the African Stonehenge, is located in Mpumalanga province in eastern South Africa. This puzzling construction of boulders weighing up to 5 tons each is about 30 meters in diameter.

Mpumalanga's megalithic areas were first described in 1891 by English traveler, explorer, and archaeologist James Theodore Bent. He estimated that there are four thousand similar stone structures in that region. However, the latest analyses indicate that there may be as many as 20 thousand of them!

Since the beginning of scientific research on these objects, fundamental questions have been asked: what was the purpose of their construction and who raised them, and when.

For a long time - until the 2nd half of the 20th century - it was assumed that they were simply remnants of the kraals, i.e. old cattle farms built by the Negroes from the tribes belonging to the Bantu group (mainly the Zulus, the population base of the Mpumalanga) who came to these lands from the north in the 18th century. Thus, no one even suspected the true age of the most famous South African megaliths today, the Calendar of Adam, let alone the role it probably played in the past.

The idea that megalithic circles were cattle pens only collapsed when researchers clearly proved that the kraals of the Bantu people were always made only from thorny bushes and not from stones.

Even later analyses of aerial photographs of Mpumalanga's vertebrae showed that a network of irrigation channels once stretched around most megalithic peoples, which earlier scientists mistakenly interpreted as traces of roads.

In turn, these dry canals ran through ancient farmland in the form of terraces. The entire region has a long history, far exceeding the period when the shepherds of the Bantu people lived there. The local Zulus people still refer to Adam's Calendar as Inzalo Y'langa, which means the Sun's Birthplace.

After Bent discovered the megaliths of Mpumalanga, for over a hundred years few people were interested in them because of the conviction of European scientists that the Negro peoples were not able to create any advanced cultures (the effect of this way of thinking was, among other things, considering the stone circles as kraals). Megalithic constructions were rediscovered for the world in 2003 by a South African pilot. Johan Heine often flew professionally over the Mpumalanga Mountains, and on that occasion saw the ruins of the structures scattered over a large area from the deck of his machine. However, he did not see the ruins at close range until one of his subordinate pilots crashed in that area. During the rescue operation, it was necessary to descend for the injured airman into a hard to reach the place at the foot of the cliff. Heine pointed out the unusual arrangement of the large stones in that place. Sometime later he returned to the site of the accident, this time having the equipment to carry out measurements. He calculated that the boulders were positioned so as to mark the four sides of the world, indicating the moments of the equinox and solstice in the sky and the place of sunrise. The pilot found the stone circle to be a prehistoric calendar - which is why the megalithic construction was called the "Calendar of Adam," the biblical first man.

When Johan Heine made his calculations public, South African enthusiasts of history puzzles became interested in the discovery. The first of them was Rodney Hale. Assuming that the reference point for the builders of the circle were the three stars of the Orion belt, he stated that the structure had to be built around 11,500 B.C. since it was then that the three stars were in the sky in a place matching the direction set by the respective stones of the circle. Then Michael Tellinger, already mentioned, was asked to explain the mystery of Adam's Calendar, who believes that traces of the oldest human civilizations should be found at the southern end of Africa.

Tellinger is an extraordinary figure. A pharmacist by education, he was a rock star in South Africa for many years. He has also produced and run television and radio programs. At the end of the 1990s, he began searching for and researching prehistoric monuments from his country. He cooperated with many scientists and presented the conclusions of his work in books that were very popular in Anglo-Saxon countries. During his research on Adam's Calendar, he was supported by Bill Hollenbach, an astronomer from the Cederberg Mountains Observatory associated with the South African Astronomical Observatory (SAAO) in Cape Town.

On June 2, 2015, the service "Ancient Origins" presented the latest analyses by Tellinger and Hollenbach, indicating an extremely distant metric of Adam's Calendar. Hollenbach, using archaeo-astronomical methods, significantly postponed the date of this megalithic construction. According to his calculations, the stone structure was built 75 thousand years ago.

A similar age of the structure was to be proved by geological analyses carried out by specialists from the Earth's past who were friends with Tellinger. A particularly important argument for such dating is the fact that some fragments of large stones forming a circle fell off due to erosion. When they were matched to the cavities, it turned out that the interaction of wind, water, and the sun crushed a layer as thick as three centimeters on the dust, which could happen in no less than 75 millennia.

In turn, the sound research carried out by Tellinger, an experienced musician and TV studio employee, showed the special acoustic properties of the arrangement of the stone blocks forming Adam's Calendar, also achieved by the piezoelectric advantages of these quartzite boulders. Thanks to both of these factors, the sound is amplified here, and then the air becomes strongly saturated with electric charges, which affects the brain of the person standing in the circle. The conditions in this construction could therefore influence people's feelings and behavior during the rituals held there, which suggests that Adam's Calendar was not only an astronomical observatory but also a cult center. Special acoustic properties are also found in Stonehenge and some other megalithic structures, as other studies have shown.

The piezoelectric properties of quartzites also speak clearly of something else: the special richness of this region. Namely, the area surrounding Adam's Calendar is rich in gold, which is why at least since the 10th century (the emergence of the nearby Negro state of Monomotapa with the center in Greater Zimbabwe), people have been boring mining shafts in the rocks there to extract the ore.

Perhaps, then, in the Mpumalanga area, scientists will someday find traces of unusually old gold mines?

And will further analysis be carried out to confirm the construction time of Adam's Calendar? After all, it is still uncertain whether we are actually dealing with a 75,000-year-old astronomical observatory, or only with an enthusiastic interpretation by Michael Tellinger, who is trying with all his strength to show that the object of his research is so old.

Temple of the Python

The factions of the San people (called Bushmen by European settlers) living in Botswana refer to the Tsodilo hills as the Mountains of the Gods or the Whispering Rocks. It is not without reason that these rocky hills in the Kalahari Desert bear these mythical names. Thousands of rock paintings have been discovered in the Tsodilo area, which the ancestors of the San created over a period of at least 10,000 years, as they considered the entire hill region sacred.

Of the Bushmen myths, several are directly related to Tsodilo. They speak of a giant serpent that created not only humans (and therefore San themselves) but also the surrounding sacred hills. He did this a long time ago, shifting the great masses of the earth with his body as he circled the Kalahari in search of water.

It was there, among the barren rocks of Tsodilo, in the fall of 2006, scientists discovered an ancient sanctuary in one of the caves, which they called the Python Temple. An international team of archaeologists, led by Dr. Sheila Coulson from the Norwegian University of Oslo, found an artifact unique in this zone of Africa, rich in rock paintings, in the cave. It is over 1.5 meters high and over 6 meters long sculpture depicting a great serpent.

"If you look closely, you can see the snake's face and eyes. It is definitely a python image," Coulson said. "Also, the place of its placement near the mouth of the cave was not chosen by chance. The sunlight enters the cave at such an angle that the rock texture of the sculpture begins to resemble snakeskin, and when it gets dark, the play od shadows give the illusion of a snake moving."

The figure was made by machining the rock protrusion in the grotto with stone tools in such a way as to give it the shape of a huge python. Near this sculpture, on one of the walls of the cave, scientists found paintings depicting an elephant and a giraffe.

When the bedrock of the rock shelter was examined, artifacts were found indicating that interesting religious rituals were being held there. There, the researchers found a pit filled with stone arrowheads. Caves of the same type have already been known from other archaeological sites in Botswana, only hundreds of kilometers from the sanctuary. All the blades found in the pit were made of red stone, and all of them showed signs of fire.

The discoverers of the Python temple assumed that these grottos were brought to the sanctuary by people not only from nearby but also distant tribes. During the religious ritual, the stone arrowheads were symbolically cleansed in a sacred fire, which probably burned in a cave. The arrowheads were then sacrificed to a deity represented by a stone serpent.

It was the physicochemical analysis of the traces of fire exposure found on the blades (or rather tar substances produced in the process of burning wood) that allowed to determine the time of performing the rituals at 70,000 years ago. This surprising finding means that the Botswana grotto can be considered the oldest so far discovered temples in the world.

"There is no indication that the cave was inhabited: we found no tools, traces of food preparation, or waste there. It follows that the place was only for rituals," Dr. Coulson explained. "This discovery indicates that our ancestors achieved a complex degree of socio-religious organization and the ability to think abstractly much earlier than we had previously assumed."

Not only this cave but the entire Tsodilo hills, teeming with prehistoric paintings, are still a sacred site of the San peoples, considered to be the ancient inhabitants of the Kalahari.

Moreover, the Bushmen are considered to be one of the oldest varieties of humanity alive to this day, and their ancestors undoubtedly inhabited all of southern and eastern Africa for at least tens of thousands of years. Therefore, it is believed that their ancestors created cave paintings in the Tsodilo Hills and a python sculpture. Does this prove that San people today still remember the ancient serpent cult? Research by anthropologists and religious scholars examining the beliefs of the Bushmen did not prove that these tribes worshiped serpents, so either this religion has long since disappeared and with it, the Python Temple has been forgotten, or ... Or the rituals are still performed, but in absolute secrecy and only by selected members of the elders of San. The latter possibility, however, is only an assumption based on the existence of secret military-religious societies among African peoples.

Now that we are stuck with presumptions, let's go even further and come up with a theory related to the python carving.

She tries to explain the origins of the cult of the great serpent, not by presenting the evolution of beliefs, but quite the opposite - based on assumptions about prehistoric fauna. It is a hypothesis by a group of cryptozoologists who believe that the sculpture from the

Python Temple depicts not a mythical creature, but an animal that lived in areas of southern Africa when the great Kalahari Desert was not yet established there.

According to these hunters of lost species, it was supposed to be a creature unknown to science, the last of which survived to this day in a friendlier region - the river and lake-rich KwaZulu-Natal province in eastern South Africa. The region is inhabited primarily by the Zulu and other Bantu tribes, but apart from them, there are also a small number of people descended from the oldest indigenous people, the San peoples. And it is in their stories that the creature referred to in their hypothesis by cryptoscientists - inkanyamba. According to local sources, the beast has a snake's body, at least 6 meters long, and its head is similar to a horse's. The ancestors of San painted images of this creature on the walls of caves for millennia because they considered the incanyamba to be an almost divine entity that had a powerful influence on the weather.

Cryptozoolologists assume that the stories of the oldest inhabitants of the KwaZulu-Natal province are not about a creature from myths, but about the most real animal - a previously unclassified species of reptile or a variety of huge eel. This assumption is to be confirmed by the occasional reports of mysterious cryptids (hypothetical animals whose existence is not confirmed by zoology) living in the lakes and rivers of the eastern part of South Africa. The only fact is that in the 1990s, a real inkanyamb fever broke out in KwaZulu-Natal. There were so many reports of this creature in the bush that in 1996 a local newspaper offered a substantial prize for a photograph of the legendary beast or other proof of its existence. Easy money amateurs ended up providing just two pictures of the inkanyamba - but both were deemed fake by journalists. So there is no confirmation of the cryptozoologists' hypothesis.

It seems, however, that the seekers of lost species unintentionally showed anthropologists and religious scholars the direction in which their mythological research should go: it would be a great achievement to prove that the legend of the inkanyamba, still alive today, comes from prehistoric beliefs in a serpent-god who was immortalized in the Temple of Python. It would be sensational to show that it is possible to survive for as many as 70 millennia in the beliefs of any people in a specific mythical thread.

The San have been living in southern Africa for 70 or even 100 millennia, but their culture was largely transformed by later arrivals - Hottentots, Bantu, and Europeans, so scientists have difficulty extracting the original elements of Bushmen beliefs from foreign accretions. However, it should be remembered that the vast region of deserts and semi-deserts in that part of Africa still hides many mysteries. Thousands of ancient rock paintings have been discovered in these sparsely populated areas that, if correctly interpreted, can tell more about the culture of the prehistoric San peoples than their heavily transformed by foreign belief influences tell us.

For now, we know so much that the ancestors of the Bushmen had rich religious images, some of which they recorded on rock carvings. We also know that the Python Temple may turn out to be the oldest sanctuary in the world, and the serpent sculpture - the oldest work of art made by humans of our species.

But can we assume that he also represents the first deity to whom Homo sapiens worshiped? This guess must remain a hypothesis, but remember that according to the story still alive among the San Kalahari tribes, mankind descended precisely from the giant python.

The serpent god who, looking for water in the desert, created the Tsodilo hills, still considered a holy place by the descendants of one of the world's oldest peoples.

The symbol from 400 millennia ago

On a fossilized shell of a freshwater mussel, there are cuts that form a pattern or sign. Some rational beings made it in a time when the species of Homo sapiens did not yet exist. Many scholars emphasize that this artifact may turn out to be a breakthrough in the history of human development.

The shell with the symbol engraved on it had been in the museum warehouse for almost 125 years. It was discovered in 1891 at the archaeological site of Trinil on the bank of the Solo River on the Indonesian island of Java, at the same time and place where the remains of the Homo erectus hominid species were first found. In 2007, the paleoanthropologist Dr. Stephen Munro from the Australian National University came across archival photographs of the unusual shell. Intrigued by the regularity of the lines visible in the photo, he found the artifact hidden in the collections and then conducted its analyses. The results were so surprising that Munro decided to commission further research on the monument in two Dutch research centers (the Free University of Amsterdam and the University of Wageningen). The dating of the marks on the shell has been confirmed, as reported by the scientific journal "Nature" at the beginning of December 2014.

"This finds changes human history," Dr. Munro said in a statement to News.com.au. "Because it suggests that Homo erectus possessed more human cognitive abilities than we previously thought. All dating records say that the notches on the shell were made between 540 and 430 thousand years ago. However, we have no idea what purpose this pattern was made. Perhaps for its creator of the genre of Homo erectus, it was an expression of an artistic act, or perhaps the symbol was made for some other reason."

Both Munro's research and independent analyzes have shown that the incisions were made on purpose and were made before the shell fossilization began, i.e. there is no way the symbol would be made at a later date. The symbol cannot, therefore, be a forgery made in the 19th century, for example.

The discovery of the sign is doubly remarkable, as the artifact comes from the Eugène Dubois collection kept at the Natural History Museum in Leiden, to which generations of paleoanthropological experts had access. And yet no one paid any attention to this monument.

But let's go back to the end of the 19th century. As I have already mentioned, the shell with stone tools and the remains of a previously unknown humanoid was discovered in 1891 - it was done by a Dutch physician, geologist, and anthropologist Eugène Dubois, who conducted archaeological work on the island of Java, then belonging to the Dutch crown. Dubois identified the fossil bones as belonging to a variety that existed much earlier than modern man and gave the creature the species name Pithecanthropus erectus, or "erect ape". It was only after many decades that scientists concluded that the creature discovered by a Dutch anthropologist was not a monkey with two legs, but an ancient form of a man. Then the name of Pitecanthrope was changed to Homo erectus.

Although the amount of data on this species increased significantly throughout the 20th century, it was only the discoveries of the last two decades that revolutionized our knowledge of Homo erectus.

It is now known that it existed the longest of all human varieties and appeared 1.7 million years ago.

Since the oldest remains of this species were discovered in Africa, it was assumed that they hailed from the Black Continent. Traces of his presence that are over a million years old have also been found in Europe, the Middle East, and Asia - from Indonesia, through China to Japan (at the same time, it died out in Africa about a million years ago).

It was in the Indonesian territories that it survived the longest because it was still there about 200,000 years ago.

It should be emphasized here that the term Homo erectus is defined by many human varieties that were once called by separate names: Sinanthropus pekinensis, the remains of which were discovered in the Zhoukoudian (Chukutien) cave near Beijing; Atlanthropus mauritanicus from northern Algeria; already described Pithecanthropus erectus from Java. Also, Homo heidelbergensis, first identified in Mauer near Heidelberg, Germany, is sometimes classified as Homo erectus, but just as often as the so-called archaic Homo sapiens sapiens, or is considered an intermediate link between the hominids from Africa and the European variety of the Neanderthals.

Homo erectus, depending on the variety, measured from 150 to over 180 centimeters in height, but was always distinguished by a slim body. The main anatomical differences between him and our species relate to the structure of the skull and teeth, and his brain was smaller than ours.

In a figurative sense, he should also be considered a prototype of the mythical Prometheus, because he was the first human species to use fire to prepare food and scare away predators. Analyzes of traces discovered in 1988 at the Swartkrans Cave in South Africa, published in early 2014 by a team of American and South African anthropologists, showed that Homo erectus used fire 1.5 million years ago. But that is not all that can be assumed about the technological and cultural advancement of this human variety.

So far, researchers have approached the possibility that hominids older than our species have a developed material culture, not to mention art. However, the dating of the symbol from the Javanese shell brings new light to two mysterious objects hundreds of thousands of years old: Venus from Berekhat Ram and Venus from Tan-Tan. Scientists have debated whether these objects are stones smoothed by the forces of nature or real artifacts.

The first of these alleged figurines, measuring just 3.5 centimeters and made of volcanic tuff, was discovered in 1981 by the Israeli archaeologist Naama Goren-Inbar of the Hebrew University in Jerusalem. The discovery was found in the Golan Heights in the Berekhat Ram area, between two layers of ash dating back over 230 millennia. Many scientists refused to accept both the age of the find and the fact that it is a sculpture of a woman. The microscopic analysis of an American scientist, Dr. Alexander Marshack, spoke for the fact that it is really an artifact almost a quarter of a million years old.

"The grooves around the neck and shoulders are deliberate tools, so undoubtedly the Israeli Venus is a statue," said Marshack.

The very date of its creation, the time when our species did not yet exist, was an obstacle for the scientific world to accept Venus from Berekhat Ram as a monument of art.

Another such object was discovered in 1999 by a German archaeologist, Dr. Lutz Fiedler. This alleged statue, named Venus of Tan-Tan, was found at an archaeological site on the northern bank of the Draa River near the city of Tan-Tan in Morocco.

Some scientists, bearing in mind the Venus of Berekhat Ram, began to consider whether to consider the find from Morocco as an anthropomorphic figure. However, they were encouraged to exercise caution by the fact that the object was excavated from a layer of earth from 500-300 millennia ago. Analyzes of the 6 cm Venus from Tan-Tan showed that it was undoubtedly made with a quartzite tool, a very hard mineral. In addition, traces of red ocher, a dye used by both our species and earlier hominids, including Homo erectus, were found on its entire surface.

Red ocher is a pigment obtained from clay-colored with hematite, iron oxide. As reported at the end of January 2012, the scientific website "PhysOrg" reported, a team of researchers discovered a deliberately stored supply of red ocher in the Maastricht-Belvedere archaeological site (the Netherlands). Importantly, the dye was produced as much as 250–200 thousand years ago and came from hematite occurring in a place several dozen kilometers away. It is not yet certain whether this stock of ocher was accumulated by Homo heidelbergensis (we remember that many scholars consider it a variant of Homo erectus) or a Neanderthal who already lived in Europe then.

As it results from the examples of finds that I have quoted, Homo erectus was the first species of hominids to intentionally use fire and to produce artifacts that can be considered monuments of religious art - statues of alleged goddesses. But that's not all, because, as we'll see, there are finds that may prove that some of the inventions so far attributed to our species were in fact the work of Homo erectus.

Material traces of the activity of various varieties of this species were discovered in very distant places because its representatives inhabited half of the world. They also made their way to islands - in the Mediterranean or the Indonesian zone, for example - which meant that they could construct rafts or even boats.

Even more surprising were the finds of houses attributed to Homo erectus. For example, in February 2000 in Japan, in the Chichibu hills (north of Tokyo), the remains of a house built half a million years ago with a roof and walls woven from branches were found, which was set on a volcanic tuff foundation. It was assumed that since Homo erectus wandered from Africa to other continents, it had to lead a nomadic life, so the house discovered in Japan should be considered a temporary, seasonal residence, and therefore having nothing to do with a permanent settlement. These assumptions, however, were challenged in 2009 by Dr. Helmut Ziegert, a lecturer at the Institute of Archeology at the University of Hamburg. According to him, Homo erectus led a sedentary lifestyle as early as 400,000 years ago and was able to create complex social structures that our species developed only in the Upper Paleolithic, i.e. not earlier than 40 000 years ago. Ziegert claims that the thousands of stone tools found in places such as Budrinna on the shores of the ancient large lake in Fezzan (southwestern Libya) and at Melka Konture on the Awash River (central Ethiopia) prove the existence of organized and long-lasting settlements of the community Homo erectus.

These settlements, inhabited by 40-50 people, were situated in an area rich in vegetation and water resources. This location allowed for generations to use local food: fruit, nuts, grass seeds, fish and birds, and land mammals approaching watering holes.

In turn, the premises for the advanced culture of Homo heidelbergensis were provided by a find from Germany. In Bilzingsleben in Thuringia, along with the 350,000-year-old remains of five individuals of this species, the remains of a camp were found - although, in reality, it was not an ordinary camp, but a settlement with a complex infrastructure. Three of its features attract attention.

First of all, the shore of the lake, on which the village was situated, was paved with stone paving several dozen meters long, as if there was a boat dock there or the local community wanted to protect itself against flooding of their homes.

Secondly, there were three fixed campfires in the center of the settlement, which suggests that there were strictly defined group activities, probably manufacturing, requiring the use of fire.

Third, there were several production zones within the village for quartering animals, stone processing, toolmaking, etc. This means that the people living there used the division of labor, and thus their group was functionally and socially diverse.

Archaeologists have also found several stone tools in Bilzingsleben that have one thing in common with a Javanese shell: geometric signs are carved on them. Scientists, however, quickly forgot about these supposed symbols, because they did not fit the image of a species living in such a distant era.

In addition to these extraordinary inventions, structures, and objects used by different varieties of Homo erectus, mention should also be made of the revolutionary tools for hunting and combat.

This species developed a method of depositing a stone blade on a wooden shaft, which allowed them to construct a spear. He also invented a javelin thrower, a simple hand-held device that allowed for much further ejection of a long projectile attached to it.

This is evidenced by finds from a lignite mine near Hanover in the late 1990s. There, archaeologists unearthed the remains of the 400,000-year-old Homo heidelbergensis camp, in which three wooden javelins and a thrower were discovered, which survived in good condition thanks to the fact that they were preserved in peat.

It was found that the oldest wooden tools known to science were carved from young trees and were perfectly balanced. The launcher accompanying the javelins allowed them to be thrown at a distance of at least 50 meters.

Probably Homo erectus was the first hominid to extract the raw materials needed to make tools by mining. In May 2004, Israeli specialists from the Weizmann Research Institute in Rehovot, analyzing finds from the northern part of Israel, determined that over 300,000 years ago Homo erectus bored shafts in order to extract high-quality flint from a depth of several dozen meters. According to the discoverers, this proves a much more advanced than previously assumed level of the social development of this species. The construction of deep mines is a complex activity requiring specialized teams of miners of various specialties and the "managerial staff" able to manage them.

And in the case of the Israeli discoveries, we are dealing with an even more complex process, because the extracted flint was later processed and made into tools that are found at sites located hundreds of kilometers away.

And for all of this, other specialists were needed: craftsmen who made tools and traders to distribute goods. The whole process was probably supervised by the administrators and the armed men who guarded them, otherwise, it would be difficult to maintain control over the multi-stage sphere of activity, such as mining, processing, and trade in flint.

If the guesses of Israeli scholars proved to be true, we would be dealing with a real breakthrough in research into the formation of social structures in such distant periods of human history.

As this short review shows, the discoveries of the last dozen or so years have revolutionized our knowledge of the hominid once known as Pitekanthrope. They show him as an inventor and a man who was able to create communities with a high degree of complexity. Therefore, each subsequent discovery of Homo erectus is valuable, because time is not kind to artifacts from such a distant era. Most of them have been irretrievably lost, and today it is even difficult to determine where to look for the remains of ancient hominids, their tools remains of settlements, because the climate has changed many times since then, and with it, the land and coastline have changed many times.

That is why tools from hundreds of thousands of years ago are so unique and valuable for science. Even more valuable are the monuments interpreted by some researchers as works of art Homo erectus, and rejected by others, because they destroy the established picture of human history.

It is not a coincidence that for over a century of dust, layer by layer covered a symbol carved over 400,000 years ago on a shell from Java.

Taking all this into account, the question arises, how many other equally "inconvenient" artifacts are hidden in museum warehouses?

How are humans different from animals? Until recently, it was assumed that the ability to think abstractly, a tendency to mystical experiences, including religious experiences, as well as the ability to use tools. However, puzzling objects discovered by scientists in recent decades question these assumptions and indicate that the animal kingdom has surprising secrets.

Chimpanzee deity

Adult chimpanzees bring gifts to an old jungle tree and play the trunk like a drum. Young chimpanzees, on the other hand, carry bits of branches with them. They do not part with them even when getting food or while playing. During sleep, they take the sticks to their lair and even prepare a separate place for them to rest at night.

Chimpanzees use many types of tools, which they make themselves from branches or stones, but - importantly - the sticks of young females are never used for food or to fight other monkeys. The sticks are worn until the female gives birth to her first child, so sometimes caring for a piece of wood takes several years. What then chimpanzees do with items they no longer need, scientists have not yet established. Maybe they abandon them, or maybe place them in some specific place.

When assessing such behavior of monkeys, it is difficult not to classify the processed branches as artifacts, even though the concept is reserved for the products of the hand and thoughts of man. In this case, however, it is necessary to loosen the corset of rules, the more so because it is the effect of the actions of the humanoid creatures most closely related to our species.

The habit of wearing sticks by young females was discovered by American primatologists, Dr. Sonya M. Kahlenberg and Dr.

Richard W. Wrangham of Harvard University, who spent 14 years studying the behavior of wild chimpanzees in the Kibale National Park in Uganda. The findings of the pair of researchers were published on December 20, 2010, in the information service "Discovery News".

Kahlenberg and Wrangham have observed over a hundred cases of young females taking care of sticks specially prepared for this purpose. Presenting their findings, American anthropologists emphasized that they spent several years collecting evidence because they were aware of the sensationalism of their discovery. The uniqueness of these discoveries lies in the fact that never before have scientists dealt with wild animals that would create mysterious objects that would perform some particularly important function in their lives.

So what are these deliberately prepared pieces of branches worn for years, if they do not serve as tools? According to Kahlenberg and Wrangham, chimpanzee sticks are simplified "dolls" used to train maternal behavior. The fact that young male chimpanzees do not wear such poles, scientists explain by gender preferences. However, this interpretation of the behavior of monkeys and the objects made by them has weaknesses. It has not been established that young male chimpanzees play with counterparts of toy soldiers or toy cars. In general, there is no reason to translate the behavior of human children operating in modern societies into the behavior of young chimpanzees of either sex. It does not take an expert in anthropology in order to see that such an attempt to explain the mystery mixes relatively new human cultural patterns with still poorly understood monkey behavior. Importantly, if we look at other discoveries about the behavior of chimpanzees, then the role of the puzzling sticks can be interpreted differently.

At this point, let us ask whether it is possible that animals have a spiritual life?

This may be indicated by the results of studies by many scientists who have indicated symptoms of chimpanzee behavior similar to those previously observed only in humans. In turn, we know that the origins of religion are based on relationships with the elements. Could it be the same for chimpanzees?

Let's start with the terrible element - fire. All animals run away from him. Well, almost all of them, not chimpanzees.

When a savannah is burning, the herd's chief male comes close to the flames, stares at them, and then begins to dance, making unusual sounds unique to this situation. Some scientists, including Jane Goodall, the most famous primate researcher, have witnessed similar behavior in monkeys, but rather related to the water element.

There was a long scientific disagreement over whether it was about typical chimpanzee behavior or about meaningless incidents. Only the results of research conducted in Senegal by Dr. Jill Preutz, an anthropologist at the State University of Iowa, proved that when chimpanzees see fire, not only do not show fear of the element but are very interested in it. Not only that - they remain in a specific relationship with fire.

Preutz collected material on chimpanzee fire behavior from early March to late April 2006, the season of seasonal wildfires in the Senegalese savannah.

Then she conducted further analyzes, the results of which were published in December 2009 in the scientific journal "American Journal of Physical Anthropology". The conclusions turned out to be surprising also in the sense that on their basis it can be assumed that chimpanzees perceive the world not only in the material sphere but also in the spiritual sphere, including those related to beliefs.

"Even though the savannah fires are violent, the chimpanzees watching the fire showed no fear, just interest," explained Dr. Preutz in a statement for EurekAlert in December 2009. "It has already been documented that when a storm hits, the dominant male in the group performs a rain dance, a behavior common to chimpanzees from many parts of Africa. Also in the face of fire, the dominant male performs a similar dance, which is why I called it the fire dance. The behavior of such males clearly showed that the dance is aimed at the fire. While they danced, the males made noises I had never heard before.

Preutz explains this behavior of monkeys according to an anthropological interpretation: chimpanzees have evolved intellectually a stage that human ancestors reached about two million years ago, that is, just before they began to "tame" fire and use it to roast meat. The chimpanzees thus have reached the 'conceptualization stage of fire', an understanding of how this element behaves under different conditions, making it possible to predict the risks of fire spread and the benefits of using it. This interpretation, however, does not explain the reason for the dances and chants performed "for" the fire by the pack leader.

The dominant male is the most important figure in the chimpanzee community. He has serious responsibilities - he maintains order in the group and leads it when it is necessary to defend territory. Leader, chief. But considering his behavior in contact with fire - maybe also a priest? He is the strongest and has the highest position in the hierarchy, so only he can come face to face with the deity to dance for him. And sing.

Dr. Preutz mentioned that the chimpanzees made noises while they danced that she had never heard before. Perhaps these are special, ritual songs directed to the fire understood as a living and powerful being. We don't know what the dominant male sings to the fire, however, because research into chimpanzee language is in its infancy.

Is fire one of the chimpanzee deities, and the leader of the herd is a priest of the destructive element? This is not known, but the behavior of these animals towards fire or rain seems to indicate that they are no stranger to the supernatural realm.

As some anthropologists suspect, chimpanzees enter into an undefined "relationship" with higher forces, another example of which is the practice of young females carrying and caring for branches. These sticks do not have to be "dolls" that prepare females for motherhood. They can be, after all, a symbolic representation of some fertility goddesses who look after chimpanzees so that after reaching adulthood they would be fertile and could become mothers.

Do the dances of fire, chants "for" the rain, and the alleged cult of fertility end the premises that may indicate that chimpanzees profess some original religion and perform rituals in honor of their deities? Well no. Many researchers have long pointed out that Charles Darwin, presenting his views on evolution, emphasized that the differences between species are graded, resembling shades of gray rather than sharp divisions into black and white. And if science has accepted the principle of continuity in the field of anatomy and physiology, why should it not also apply to the emotional, moral, and spiritual spheres? And not only in humans but also in animals? More and more scientists assume that animals are capable of experiencing metaphysics, gaining spiritual experiences. The basis for such a conclusion is - paradoxically - studies on humans, which have shown that spiritual experiences are related to the most recently developed part of the human brain, the structures of which also exist in the brains of animals.

"Since only humans are capable of describing the richness of their spiritual experiences in complex language, it is unlikely that we will ever know what the subjective experiences of animals are," Dr. Kevin Nelson told Discovery News on October 8, 2010. - Despite this limitation, the conclusion is still valid that since most of the primary areas of our brain are related to spirituality, animals should also be capable of this type of experience.

I quoted Nelson's opinion, not by accident. The scientist is a lecturer in neurology at Kentucky State University and the author of the book "The Spiritual Doorway in the Brain", published in January 2011, in which he collected the results of research conducted by him for thirty years.

The work of Dr. Nelson and other neuroscientists dealing with this topic has established that, for example, experiences such as a spiritual out of the body (the so-called OBE phenomenon) can be triggered in humans by stimulating this primary part of the brain as it regulates different states of consciousness.

"We already know that if we interfere with the area of the brain that regulates a person's visual experience, sense of movement and orientation in the Earth's gravitational field, we will induce OBE in the subject's consciousness as effectively as using a mechanical switch," explained Nelson. There is absolutely no reason to believe otherwise for other primates or even for a dog or a cat.

According to his research, mammals can have dying experiences similar to those described by some humans. This is especially true for the phenomena described as seeing light and moving towards it through a tunnel.

"The end-tunnel light phenomenon is caused by the drop in blood pressure in the eye that occurs with fainting or cardiac arrest," said Nelson. "There is no reason to believe that animals perceive this state differently. Another thing is what they see in the tunnel."

Science still cannot explain precisely why people experience mystical experiences. Are they the cause of the body's behavior discovered by scientists, or are they the result of it? The fact is that researchers are already able to imitate such experiences, disrupting the operation of the appropriate part of the brain. However, they still do not know how such sensations take place outside the laboratory and what triggers them "in nature". And yet these issues are very important because mystical experiences are the basis of all religions.

So one can ask if the animals also have such experiences, and thus - can some animals have their own religions? This is indicated in chimpanzees by fire dancing and singing in the rain and the wearing of a stick by young females. And as we'll see later, these kinds of chimpanzee behaviors are even more complex. But what does an expert in neuroscience think?

"Mystical sensations, and therefore fleeting moments of feeling a supernatural mystery and amazement with this mystery arise in man within the limbic system," said, Dr. Nelson. "Since animals such as primates, horses, dogs, and cats also share similar brain structures, it is possible that they too have mystical experiences, including a sense of spiritual unity with other members of their species, and perhaps even with all creatures."

Dr. Marc Bekoff, a retired professor of ecology and evolutionary biology at Colorado State University at Boulder, goes much further in his views. In an article published in early October 2010 on the website "Wildlife", he argued that animals have spiritual experiences comparable to humans. As he wrote: "these experiences bring intangible values to animals and provoke them to introspection".

Bekoff, like the famous primate researcher Jane Goodall, watched chimpanzees dance by the waterfalls that formed in the African highlands after heavy rains. According to both of these scholars, some of the chimpanzees seemed to enter a trance while dancing, similar to the state that people who participate in religious rituals sometimes get. Goodall, in his study on primate spirituality in The Encyclopedia of Religion and Nature, seriously wondered whether chimpanzee dances were religious behavior and, consequently, whether these animals were "precursors" to devising religious rituals.

An American anthropologist described the behavior of a chimpanzee approaching the waterfall with its hair bristled, which was very agitated. As Goodall wrote: "he stands upright, sways rhythmically, shifts from foot to foot, splashes the water with his feet in the shallow puddles that form near the waterfall. Such a Waterfall Dance can last ten to fifteen minutes." According to the findings of this researcher, chimpanzees perform similar dances also at the beginning of heavy rains (which I already wrote about when talking about Dr. Preutz's discoveries) and when violent gusts of wind come.

The researcher, therefore, pondered the question of whether chimpanzees, while dancing among the "wild elements," were practicing a ritual belonging to some form of animistic religion. Do they worship the flood from heaven, thunder, and lightning? Isn't it possible that these chimpanzee performances are stimulated by feelings similar to the human fear of God? It is worth spending years of research to discover what animals see and feel when they look at the stars. ' Supplementing these thoughts, it is worth quoting a sentence from an article by Dr. Bekoff, published in the specialist journal "Psychology Today":

"Perhaps many species of animals participate in such ceremonies, but we are not lucky enough to see it. For now, let's leave the door open to the thought that animals can be spirit-experienced beings."

At the outset, I mentioned pieces of branches carried by young chimpanzees. They can be considered artifacts of mystical importance, the most primal form of representing the fertility deity. We found many manifestations of the spirituality of primates still classified as animals. Science has faced a great challenge to explain these phenomena as more and more studies indicate that animals can have mystical experiences. Maybe even religious, because the behavior of the African jungle chimpanzees appears to be animistic rituals similar to rituals proper to primitive human beliefs.

So let's say it bluntly that chimpanzees perform rituals related to fire, rain, storm, wind, and fertility worship. So if these apes are really close to mystical feelings, they should also understand what death is. They should perceive the moment of transition between the material and the spiritual spheres. And it is so because new research on chimpanzees has revealed that these animals give "funerals" to their loved ones, and therefore feel or even understand the concept of death.

One of the chimpanzee funerals was held in October 2009 at the Sanaga-Yong Chimpanzee Rescue Center in Cameroon. Dorothy the chimpanzee died there of a heart attack.

As reported on October 28, 2009, in the British magazine "The Mirror", members of the monkey herd organized a farewell ceremony for her deceased companion. The entire chimpanzee colony mourned its dead - 25 adult monkeys stood silently by the wire-mesh fence, accompanying Dorothy on the last journey. The chimpanzee caravan was not very sophisticated, typical of African realities, namely, the body of Dorothy was placed by Sanaga-Yong personnel on a wheelbarrow. The chimpanzees were kept quiet and standing in a tight group while taking the body and then transporting it outside the facility. Such behavior, considered purely human, shows that the difference in the perception of strong emotions by us and by monkeys, after all, considered animals, is either small or even non-existent.

Funeral ceremonies performed by wild chimpanzees are harder to observe. However, a group of researchers led by Dr. Dora Biro from the University of Oxford managed to do so. According to data published in April 2010 in the journal New Scientist, Oxford scientists for many years analyzed the behavior of a herd of chimpanzees living in the jungles of Southeast Guinea. They also observed their reactions to the death of loved ones. One such case happened in 2003. Jimato, a year-old male, then died. His mother carried the baby's body with her for over a month. The mother of Veve, 2.5 years old, who died in 2010, behaved in a similar way.

The chimpanzee carried her daughter's body for almost three weeks. After this period of alleged mourning, "the bodies of the deceased were abandoned by their mothers," as reported in the study report.

But was the body abandoned? The researchers found that in both cases, the chimpanzees left their dead babies among the dry, sun-warmed rocks. These were places where the prevailing conditions quickly led to the natural mummification of bodies.

Scientists are cautious that this was a deliberate act, but if it was, it is a funeral that was deliberately mummified by natural conditions such as heat and dry air. There are peoples who buried their dead in this way, and therefore mummified them in the sun and then left them in the air or hid them in tombs.

The results of studies by other British researchers also indicate that chimpanzees understand the concept of death and celebrate mourning in the form of rituals. A team led by psychologist Dr. James Anderson of the University of Stirling in Scotland has conducted and continues to conduct long-term research into the behavior of primates. As reported on May 2, 2010, by the WENN agency, scientists from Stirling filmed the funeral behavior of a group of chimpanzees. An elderly female named Pansy died then.

Her daughter stayed with the body all night. The next morning, the rest of the herd cleaned Pansy's body. They were petting the deceased. At the same time, they avoided the place where the chimpanzees died. For a few days after the death of their companion, the monkeys behaved very calmly, just as if they were in mourning.

"Science has several areas to distinguish humans from animals," Dr. Anderson explained. "Today we have more and more evidence that these differences are not as clear as it was once believed. The awareness of death is another psychological phenomenon that we can also observe in animals."

A team of Scottish scientists has been working with this group of chimpanzees for several years, so the next statement by Anderson, published in April 2010 in New Scientist, is significant:

"Our research shows that chimpanzees not only understand the concept of death but also know how to deal with the loss of loved ones. For this purpose, they perform some kind of ritual."

As the examples cited show, chimpanzees consciously bid farewell to the dead. Is it just about the consequences of the social life of monkeys, or is it something more complex and distant from "animal nature"?

New scientific findings indicate that the closely related primates understand the meaning of death, and therefore undergo a period of mourning and perform the rituals associated with it.

The question remains whether this means consciously sending the souls of loved ones to another world, where animistic deities await them.

The above findings and hypotheses indicate that many chimpanzee behaviors may indicate rituals related to shamanism or animism. And are the sticks mentioned at the beginning, carefully selected and carefully cared for by young chimpanzees, really dolls preparing for motherhood, as Sonya Kahlenberg and Richard Wrangham assumed? Or maybe, as I have already written, they should be considered as primitive images of the female deity of fertility? This possibility is definitely worth considering. The more so because scientists have discovered a chimpanzee temple.

A team of biologists led by Dr. Laura Kehoe from the Humboldt University in Berlin, working in the Republic of Guinea, made a discovery that demolished the common perception of the uniqueness of man as the only creature that creates places of worship. As reported by New Scientist on March 4, 2016, Dr.

Kehoe's group of scientists has found evidence that Guinean wild chimpanzees treat one of the trees of the tropical jungle in a special way.

In the course of the fieldwork, Dr. Kehoe noticed a tree in the daily migrations of a herd of chimpanzees. The bark of this old tree was damaged, scratched, or chipped in many places. Even more surprising was what was in the great hollow situated at ground level. There was a pile of round stones.

Scientists have set traps around the tree - cameras activated by motion. After a few weeks, the researchers got acquainted with the digital record, which revealed something they sensed, but were afraid to say it out loud beforehand. Namely, chimpanzees approached the tree with specially selected or processed stones. They were all round and of similar size. The monkeys first hit the trunk with stones and then put them into the hollow.

A tree with a large hollow at its base resonates sound, so it is a drum made by nature. The monkeys hitting the trunk with stones create a loud bang. Dr. Kehoe's team assumes that it is a kind of communication medium, but symbolic.

Is it a "sacred tree" used by chimpanzees to perform some ritual? Probably yes. Simply put, it's a temple.

"Perhaps this is evidence that the chimpanzees have created a sanctuary of sorts," explains Kehoe. "Taking into account all the findings, the following hypothesis can be made. Well, the sacred tree makes sounds, and on the altar, which is a hollow, chimpanzees lay votive offerings, or stones, with which they produce sound." Of course, there is no clear evidence that chimpanzees believe in a particular deity and that this is a ritual by which they worship that deity. However, we have learned that chimpanzees display extremely rich behaviors that we can associate with early beliefs.

Dr. Kehoe's team was joined by Dr. Jill Preutz of Iowa State University, whose research on "fire dances" by chimpanzees I mentioned earlier. No wonder that this scholar also approaches the alleged sacred tree with an open mind.

"Male chimpanzees have repeatedly been reported to hit tree trunks with thick branches or stones, making loud noises that go beyond the standard chimpanzee shouts," Dr. Preutz told New Scientist in early March 2016. - It seems to be a tradition in some groups. However, this is the first time that chimpanzees have been shown to store stones in a hollow tree. If it fits the definition of a proto-ritual, I have no problem with that.

The rituals at the tree and the offering of votive offerings in it are another behavior of chimpanzees filmed by trap cameras showing that these monkeys have developed religious life. When we combine a chimpanzee temple with their dances of fire, rain, images of the fertility deity, and finally with funerals, we get a picture of human communities from tens of thousands of years ago. The concept of an artifact, previously reserved for the products of human hands, needs to be remodeled. For what are the pieces of branches worn by young chimpanzees and round stones deposited in the hollow of the sacred tree, if not artifacts? The creative, or rather technical, the activity of chimpanzees does not end there, as will be discussed in more detail in the next chapter.

Inhuman tools

In May 2002, German archaeologists from the Max Planck Institute for Evolutionary Anthropology and American researchers at George Washington University stumbled across the jungles of East Africa where they discovered many stone tools. The monuments were similar to the oldest known tools made millions of years ago by human ancestors, but they turned out to be hiding a surprising secret.

According to the current scientific views, the first being able to produce simple tools was Homo habilis, our relative walking on two legs in an upright position, considered to be the first link of species referred to as Homo, ie "human". However, the tools found by the Germans and Americans, although remarkably similar to the artifacts he produced, were at most a few hundred years old, and some of them looked as if they had been made yesterday. Scientists began to ask themselves whether they had found a trace of an extremely primitive tribe in tropical forests, which until recently lived, and maybe still lives in those areas.

First, it was found that toolmakers used them to crack nuts. Lots of hewn stones have been found both on the surface of the earth and at shallow depths that serve as "hammers". 40 kilograms of nutshells of various species were also collected. As the researchers rightly guessed, they were dealing with a "specialist workshop" where both tools intended solely for cracking nuts were manufactured and the nuts were split. Even a cursory examination of the site revealed that the site was still used for the preparation of tools and food since the freshest broken shells were only a few days old!

When the scientists realized this, they quickly collected some tools and some shells as samples and left the site of discovery, trying to mask the traces of their activities. However, they left the motion sensor-activated trap cameras.

Scientists already suspected who the user of the enigmatic workshop might be, but wanted compelling evidence. Soon, the photo material dispelled all doubts - it was about a local herd of chimpanzees. Their observation showed that monkeys first select stones that are appropriate for shape and size, and then use them to make hammers which they use to crack nuts. Researchers were thus faced with a whole deliberate process of making tools and using them in food preparation.

The last stage was that the monkeys put the nut between several branches, which were hit with a stone from above until the stuck walnut broke. As it resulted from further long-term observations carried out by scientists, older individuals taught this activity to young individuals, which took several years to learn.

This discovery proves that monkeys can make tools and use them intentionally and long-term (not just once). And it has probably been happening for millions of years, and they pass this knowledge down from generation to generation. As the lineages of monkeys and humans split around 14 million years ago, scientists began to suspect that the manufacture of stone tools intended solely for the cracking of nuts is the earliest hominid manufacturing and cultural activity.

This, in turn, leads to the conclusion that the ability to create and use simple tools may be much older than previously assumed, because if both hominid lines had it, it is reasonable to suppose that it happened even when they formed one species, i.e. a dozen or so species. million years ago. If this guess proved to be true, it would imply that it was not Homo habilis who invented the stone tools.

We already know much more about the inventiveness of chimpanzees living in the wild. In October 2004, scientists observed a herd in the Nouabalé-Ndoki National Park forest in the Republic of Congo, the members of which were making other tools. Monkeys used them to catch termites, which are valuable high-protein food. They used large, heavy sticks to punch holes in the termite mounds. Then they took previously prepared long, thin twigs, which scientists called fishing rods, and then gouged out insects with them. To search deeper parts of the termite, chimpanzees used other spade-like sticks to help pierce the nest and select insects. According to scientists working under the auspices of the New York Wildlife Conservation Society and the Congo authorities, the chimpanzees studied had never had any contact with humans, so they could not "pull" such behavior from them. So it follows that they themselves invented the tools they needed.

This is not the end. In February 2007, a team led by Dr. Jill Preutz, a primatologist at Iowa State University, the one who described "fire dancing" to chimpanzees, observed that chimpanzees living in Senegal are making weapons and using them to hunt other mammals. They made spears from branches and used them to hunt small galago monkeys. The hunts were as follows: a chimpanzee stuck a spear into a pit where some galago rested during the day, and thus killed or immobilized the victim. However, according to Dr. Preutz, chimpanzees are not very good hunters, with one successful hunt for an average of 22 attempts.

Later studies of other herds of chimpanzees - both in East and West Africa - have shown that the use of simple tools is much more widespread among this species of monkeys than was believed until the end of the 20th century.

What about tool handling by other ape species?

It was once assumed that gorillas, as exceptionally strong creatures and committed vegetarians, do not use tools because they do not feel such a need. Admittedly, it was observed that these monkeys use objects, helping themselves with food, for example, but they were always taken in captivity, so researchers thought that they had to imitate human behavior. However, in early October 2005, two female gorillas were photographed living in the wild, using branches to get across the river and its boggy banks.

A group of scientists from the Wildlife Conservation Society and the German Max Planck Institute has been observing gorillas living in the northern part of the Congo Republic for 10 years. Researchers had seen these monkeys twice using simple tools, but it wasn't until 2005 that they documented gorilla behavior in photographs.

Two females, photographed by scientists, intended to cross the marshy area in the Mbeli Bai forest. The muddy ground stopped them, so they looked for the appropriate tools. The first monkey found a long strong branch. She used this pole to check the depth of the water where she wanted to get to the other side. Then the same branch served as the gorilla's staff.

The second monkey came up with a different strategy: it found a large branch, placed it on the marshy ground, and walked dry foot over an improvised footbridge.

"This is an amazing discovery," said Dr. Thomas Breuer in charge of the research. "It will allow us to better understand the evolution of our species and animal skills."

Other research suggests that perhaps all primates, not just chimpanzees and gorillas, make useful items to make their lives easier. In January 2003, American researchers led by Dr. Carel van Schaik from Duke University in Durham, North Carolina, found a herd of orangutans on the Indonesian island of Sumatra using a variety of tools. The social culture of these great apes also turned out to be extremely complex. For example, before going to bed, members of the herd exchanged hugs and glances, as if they were saying "good night". Durham researchers found that the creation and use of simple tools were common and ordinary for the subjects of the observed group. They were used not only to obtain food or chase away insects with leafy branches but also for such interesting activities as cleaning one's teeth with small sticks. Added to this are unusual behaviors and tools. For example, some orangutans rolled large leaves into their proboscis and made terrible sounds with them. Scientists have found no rational cause for this behavior - apart from one: play.

Above all, however, orangutans used many tools made of wood and stone. They were helpful in removing insects from under the bark of trees or extracting the flesh of fruit and nuts from their hard shells. The toolkit was similar to that of chimpanzees previously described. It should be emphasized that the studied orangutans live in the depths of the rainforest, where they have no contact with humans, so it can be assumed that they invented the tools they use on their own.

Let's move beyond the great apes because scientists have found that they are not the only ones who use tools. In mid-December

2004, Dr. Antonio C. de Moura from Brazil and Phillis C. Lee from Scotland, anthropologists working as part of a research project sponsored by the University of Cambridge, made public the results of their many years of research conducted in the Serra de Capivara National Park in Northeast Brazil.

A pair of scientists have established that the crested capuchin females living there use tools, including even stone knives, on a daily basis. These little monkeys pick pebbles from the ground, and then select them in terms of size and shape for the type of activity they plan to perform with them. Tools are used by capuchins to dig up edible rhizomes and roots, to extract insects from under the bark of trees, and to split the shells of seeds and nuts.

In addition, monkeys use properly selected stones as knives. They use them to cut the collected plants and hunted insects, lizards, and small rodents into pieces that are easier to eat.

Capuchins also choose twigs suitable for rummaging in rock crevices and hollows to extract insects and honey from there. Scientists have found that the use of tools by capuchins must be a skill mastered a long time ago because these monkeys mainly eat food that would be unavailable to them without the use of properly selected tools.

It is now known that you don't have to be human to invent and manufacture tools. It is enough that - like a monkey - you have two hands and ten fingers. However, is it a necessary condition?

It turns out that no, because you don't need hands to make a useful tool. All you need is a mouth and ... hooves. It turns out that in the group of species that use tools there are also pigs. On October 8, 2019, it was reported by the "CNN" service, and the case was presented in more detail by the specialist journal "Mammalian Biology". A team of scientists led by environmental ecologist Dr.

Meredith Root-Bernstein of the Musée de l'Homme in Paris investigated the behavior of dwarf pigs living in the wild in the jungle of the Philippines. Scientists found that these animals search for pieces of the bark of the right size and thickness and, holding them in their mouths, use them as a shovel to dig a hole in the ground for the cubs' den. This is not the end of the "tool" talents of Filipino pigs.

They live in the jungle near human settlements and often forage in the fields. So people, to protect vegetable crops, build fences with "electric shepherds". Scientists led by Dr. Root-Bernstein have found that some Filipino pigs manage these safeguards. Namely, a pig, intending to eat a man-grown vegetable, searches for a stone. He grabs it in the mouth and throws it at the steel mesh fence to see if it's dangerous, and therefore energized.

Another, even more extreme, example of instrument-wielding mammals is the dolphins that live in the water. They use sea sponges as tools to aid in hunting fish.

According to biologists who study these mammals at the University of New South Wales, Australia, the making and use of tools by dolphins is a mother-to-daughter cultural behavior, as shown by sightings of bottlenose dolphins living in Shark Bay on Australia's west coast. The bottlenose dolphins there employ a variety of hunting tactics, one of which involves the handling of tools that the animals make themselves. They break off a piece of the sponge and then, holding it in their mouth, use it to comb through the vegetation on the bottom, which helps them find smaller animals to feed them. Biologists have established that such tools are used almost exclusively by females who pass this skill on to their daughters.

The same researchers in the same body of water in August 2011 found dolphins who invented a different tool.

They were large shells with which bottlenose scattered fish trying to hide among the unevenness of the bottom covered with coral.

Dolphins used both the shells from which they first ate the mollusk, as well as those already empty, lying on the bottom. Experts from the University of New South Wales emphasize that this is undisputed evidence of the existence of material culture among marine mammals. And - importantly - not the only one.

In March 2008, scientists from Great Britain and Brazil studied the behavior of South American freshwater boto or Amazonian inii - Inia geoffrensis; animals sometimes called river dolphins. These scientists discovered that these mammals also use a variety of objects, not to obtain food, but to ... mate. Boto lives in the Amazon and the Orinoco. The average individual weighs about 100 kilograms and grows up to three meters in length. Amazonian inia has a special beauty because although it looks a bit like a dolphin, its skin has an unusual gray-pink color. Scientists have found that during mating season, males of this species use stones, pieces of wood, lumps of clay, and clumps of grass found in the river to impress females and to scare off other suitors. Such items are an important element of the mating ritual because they help to attract the attention of nearby females when a male inia chooses to compete. When a few of them, curious, swim towards the male, he shows them what he has found interesting and starts hitting the object in his mouth against the surface of the water. Sometimes the boto raises the tension in this show. It leaves the object at the bottom, floats to the surface, then dives quickly, grabs the object, emerges vertically from the depths, and presents the prey held in its mouth.

The more the female likes the object and the display of the dexterity of its extraction, the more the male gains in their eyes and can count on a love adventure and even a permanent relationship.

Dolphins are specific animals. They resemble primates, including us humans, in many ways. They can act both individually and in a group, and they also have social habits. Recent studies show that they use complicated languages and even understand the fact of death and mourn their dead tribesmen. Perhaps the mystery of their development lies in the fact that the brain of a bottlenose dolphin weighs an average of 200 grams more than the average human brain, and that the cortex of these aquatic mammals contains more ganglia than our cortex? On the other hand, the previously presented capuchins who use stone knives have brains as small as a domestic cat, so it turns out that neither a great dolphin's brain nor skillful monkey hands are needed to make tools. It is enough to have a beak, as evidenced by the achievements of corvids. The birds in this family are intelligent. They can make tools, and what's more, they know how to improvise in their activities, because they often use previously unknown materials to make tools that enable them to get to the food.

"The instinctive tendency of corvids to seek solutions involving the use of tools is unusual in the behavior of animals. Some of their abilities mean that humans are less distant from a raven or a crow than an adult chimpanzee in this respect," Professor Alex Kacelnik, an Argentine zoologist teaching at Oxford University, told media on February 10, 2006.

Kacelnik conducted many experiments with corvids. He discovered their possibilities by accident, observing the behavior of

Betty, a New Caledonian crow, kept in a laboratory cage. This bird grew up in the wild and was taken prisoner at the age of two. Professor Kacelnik published the first results of research on the behavior of Betty in August 2002. He found that this crow - despite the lack of training and without human help - can choose a suitably flexible but durable piece of wire, then bend it into a hook and use it to pull out food hidden by the researcher in a hard-to-reach place. Betty, in order to get to the chopped meat hidden in the container, made a hook from a material unavailable in nature, but known to this bird, because the bars in its cage were made of metal. The crow, holding the hook in its beak, inserted it into a tube attached by scientists in a container with food. With such a tool she could reach the meat and pull it out piece by piece.

Experiments have shown that this common bird not only associated cause and effect relationships but also can make a tool appropriate to achieve the goal.

"We had to find out that it was not just luck. We repeated the tests. The animal managed to achieve the goal in nine out of ten attempts," explained Kacelnik. "People assume primates are the most intelligent because they are the closest to us. However, as it turned out, Betty seems to be on par with primates in this regard."

The studied bird belonged to the species Corvus moneduloides (bearded crow), found only in New Caledonia and the surrounding islands, but closely related to our crows. In each subsequent experiment, Betty first tried to reach for the food with a straight piece of wire. Only when it turned out that nothing could do that, did she perfect the tool. To bend the wire and thus obtain a hook, she wedged one end of the wire in a piece of plaster securing the base of the food tube to the plastic base or held the wire at some distance from the tube with her foot. On the one hand, this proves the bird's great ingenuity and intelligence, on the other, however, it says that it did not remember the lessons from previous experiences, and therefore did not have the ability to permanently learn.

"Betty's success is based on the use of material not available in nature and the implementation of a specific task with the tool she has made," emphasized the Argentine professor. "Importantly, the crow did not have a role model."

In January 2005, Kacelnik and his team from the Department of Zoology at the University of Oxford published the results of further studies on young crows from New Caledonia, but it should be noted that these individuals were raised in captivity in the Royal Botanic Garden in Kew.

Earlier, however, Kacelnik established that other wild birds of this species use pebbles, twigs and spikes as tools. It was suspected that the young learn such skills from their parents. However, observations and tests have shown that this ability does not come from learning from older individuals, but manifests itself spontaneously between days 63 and 79 of the bird's life. It can therefore be "genetic knowledge" or the result of individual observations and experiences, which, thanks to species intelligence, lead individual individuals to solutions to the problems and challenges they encounter. So it turned out that Betty the crow is no exception - her relatives are equally intelligent.

The Corvidae family, or corvids, includes over 100 species of birds from around the world with a varied lifestyle. Their high intelligence is confirmed by more and more studies, including analyzes of an international group of biologists from Great Britain, New Zealand, Spain, and Canada. The results of this team's work, published in mid-March 2005, clearly showed that birds with large brains in relation to their body weight adapt more easily to the new environment and learn to find food faster. Data on as many as two thousand species of birds were analyzed - corvids were the best in each field. An interesting example of their intelligence is the behavior of Japanese ravens who use ... cars as nutcrackers.

The common ravens (Corvus corax) living in Japan leave their nuts on the roadways of busy intersections and wait for their delicacy to be broken by a car.

Ravens were even observed waiting for the red light to come on and stop the traffic to lay down the nuts and then pick them up.

The number of animal species making and using tools grows as research expands. But what Dr. Culum Brown, a professor of ecology at Australia's Macquari University, found that surprised everyone.

According to the findings of this researcher, published on December 1, 2011, on the website "PhysOrg", even fish use tools. Research on the fauna of the Great Barrier Reef off the coast of Australia by Brown's team found that common catfish use tools to obtain food. These fish pick up rock-hard pieces of the reef from the bottom and, holding the equivalent of a hammer in their mouths, break the shells of crustaceans with it. After breaking the covers, they eat the defenseless clam.

'Fish carefully select pieces of dead coral and then use them with great precision, indicating experience with such operations,' Brown explained. "This is not a hypothesis, but a statement of fact, because it is backed up by a multitude of crushed shells lying in many places on the reef where catfish feed. So far, fish have not been appreciated for their skill. However, the acquisition and use of tools are evidence of the intelligent behavior of at least one species. More time is needed to study marine fish in the wild to find out how common tool use is among them.

It is not known how and when the catfish discovered that they could get food with a piece of rock. Was it 10 years ago or a million? But this is not the end. It turns out that to become an inventor you don't even need a backbone.

Literally, because even invertebrates use tools. Dr. Julian Finn, an Australian marine biologist at the Victoria State Museum in Melbourne, studied the marine life off the coast of Indonesia. According to conclusions published in mid-December 2009 in the journal New Scientist, the local octopus of the species Amphioctopus marginatus uses tools for protection.

These small cephalopods living in tropical waters collect coconut shell halves from the seabed, which are thrown into the water by humans. Octopuses carry these shells near the rock crevices in which they live. When they go out hunting, they cover their soft body with half of their coconut-like armor. The term "they go hunting" should be taken literally, because this species has the ability to walk on the bottom on two legs. The remaining tentacles are used to capture prey.

The creating armors by Amphioctopus marginatus are the first reported case of tool use by invertebrates observed in the wild.

"This behavior of octopuses should definitely be considered an ability to acquire and handle tools," Finn said. "Coconut shells are collected on purpose and used when the octopus deems it necessary."

It can be said that the cephalopods operate on a schedule, unlike the hermit crab, which simply takes the shell of a dead mollusk and wears it as protection.

It should also be taken into account that the collection of shells by the octopus is energy-intensive for it, and it can sometimes expose the cephalopod to predator attack. However, in the long run, having an armor that can be put on when needed is very profitable for Amphioctopus marginatus.

Thanks to the discoveries of the unusual behaviors of octopus, catfish, bottlenose dolphins, and boto, another popular belief about the non-use of tools by sea and river animals have collapsed. It turned out that even creatures living in an aquatic environment can create the beginnings of the technology, because that is, after all, obtaining simple tools from the environment and then using them to gain benefits.

It was long assumed that the first tools were invented by the ancestors of modern man - and only when they began to move in an upright posture.

The new findings show that our bipedalism and our large brains do not have to be prerequisites for the ability to create and use tools. What's more, tools are not the exclusive cultural achievement of a human being, and as recent research shows, they are even quite a popular invention in the animal kingdom.

Another view that must be changed as a result of scientific progress is the assumption that man is the most intelligent creature on Earth. Research in some research centers shows that things are a bit different.

This applies not only to the manufacture and use of tools but extends to all levels of animal functioning.

As reported in December 2013 by PhysOrg, evolutionary biologists at the University of Adelaide in Australia, led by Dr. Arthur Saniotis, showed that many species of animals exhibit high intelligence, but it is very different from our intelligence.

According to Australian experts, humanity does not understand such varieties of intellect, because it has adopted the ability to create complex forms of communication and sophisticated technologies as a measure of mental development.

"Since the invention of agriculture and husbandry, and then codified the rules of life in the form of religious dictates, people have fooled themselves into being more intelligent than animals," said Saniotis. "Scientific findings prove, however, that many animals have cognitive abilities that exceed our abilities, only of a different type. So humans aren't smarter than many animal species, they're just different."

And this assumption may explain the apparently incredible abilities of animals. For example, ravens and dolphins invent various tools to make their lives easier.

Looking at the tool use problem from this point of view, even fish breaking clamshells with coral hammers does not seem strange. And even more so, chimpanzees, who run a "workshop" in the African jungle, where they have been producing stone devices for cracking nuts for generations.

Bird money

A few-year-old girl likes to spend time in her home garden. The company of birds gives her special joy. The story takes place in the New England town of Seattle on the East Coast of the United States, and these birds are not colorfully feathered parrots, but croaking and screeching crows. However, the child finds them very interesting, and it must be admitted that he has reasons for it. Why? Well, thanks to the contacts with these birds, the girl learns the rules of ... trade. He lays out food in a designated area of the garden every day, and the feathered consumers come to the feeder. The most amazing thing, however, is that they pay the bill for meals every now and then. They pay with items, for example on November 9, 2014, at 2.30 pm, they brought a small broken light bulb.

Many people enjoy it when birds appear in their garden - then they observe their behavior, listen to singing, and in winter they feed the winged company. Usually, however, these are one-sided relationships, because one can get the impression that the birds do not reciprocate our emotions, and that they use karma without reflection. However, the contacts that eight-year-old Gabi Mann from Seattle made with the crows indicate that in this case, it is a deliberately conducted barter.

The story of little Gabi was published on February 25, 2015, by the BBC News news service. The girl became friends with the birds when she was four years old. She watched crows fly into the backyard garden and was very interested in the behavior of these creatures. She started feeding the feathered guests - every day, in the same places, she scattered out a few handfuls of peanuts and some dry dog food. This composition of meals was proposed by Lisa Mann, Gabi's mother. It turned out that the woman liked the crows. And when she put a concrete drinker in the garden and made sure that there was always fresh water in it, the birds chose this gastronomic place. You could say that they were regular customers because they started to treat the feeder like a restaurant.

And, as is the case in restaurants, they did not come empty-handed (or rather, in their case, with an empty beak) to feast every day. Every now and then, they settled the meal bill by bringing in small items.

Since Gabi went to school in September 2014, the rules of her relationship with crows have expanded. In the afternoon, when the girl comes home, the birds are waiting for her, hanging out on the telephone lines. When the school bus for children stops in front of Gabi's house, the crows start croaking loudly, which can be considered a greeting. The eight-year-old enters the house, takes the food, and goes with it to the garden. The birds then flock to the lawn near the drinker, and the girl gives them peanuts and granulated dog food.

The crows eat on credit, but once every few days they settle the bill, leaving Gabi behind with something. It is not known when exactly they do this, because the child's mother discovered it only in mid-2013. Lisa Mann realized then that she was witnessing something extraordinary and approached the subject with a scientific flair. She persuaded her daughter to start to keep a record of artifacts brought by the winged guests.

"I like that Gabi loves animals and wants to share food with them," Lisa Mann told BBC News. "As you can see from the behavior of crows, not only people try to reciprocate when someone is good to them and offers them something. Before my daughter became friends with them, I didn't even realize they were near our house. I didn't pay any attention to the birds at all, but my daughter gave me a kind of transformation."

What items do crows pay for regular meals in the garden canteen? What do they consider valuable to humans? The answer is a collection of items inventoried over two years by Gabi and her mother, stored in a plastic container with compartments. Items delivered by birds are small and light, after all, they must fit in a crow's beak.

At the beginning of 2015, there were almost a hundred of these artifacts. Each was placed in a transparent box or plastic bag, labeled and classified, and placed in a separate compartment of the harvest box.

"Bird money", always put in the same place where the girl pours out the food, are various items. Pieces of brown glass from broken beer bottles, a miniature silver ball, a black button, a blue paper clip, yellow balls, shiny pebbles, a piece of once black and now faded foam for seals, a blue Lego brick, a rusty screw, beads of various colors, buttons with mother-of-pearl, metal earring. Some items are in very good condition, others are scratched and dirty. You can see that when it comes to assessing the material value of artifacts, crows and people are definitely different. All this history shows, however, that both species value the principle of reciprocity equally highly. Of course, the opinion of crows in this matter is unknown, but Gabi treats her relationship with animals very personally, and from their gifts, she especially appreciates a piece of the metal pendant with the inscription "best" on it.

The second part of the item is missing - presumably where the word "friend" was - but the girl takes the birds' intentions for granted.

The crows keep a close eye on their canteen and its staff. They can also pay bills not only with items but also with favors, as Lisa has experienced personally. Since 2013, he has been regularly photographing crows in his garden and recording their behavior and forms of interaction. She was so drawn to this matter that bird photography became her hobby.

One day, when she noticed flying specimens of other species, not crows, she began to take pictures of them. She left the garden and walked down the street. In a nearby alley, she finally took a few satisfactory photos. Unfortunately, she was so busy she lost the plastic lens cover. The next day, however, she found her lost - one of the crows brought it and placed it on the rim of a concrete drinker, where the birds usually paid their meal bills.

"I'm sure it was intentional," Lisa assured. - The crows are watching us all the time. One of them saw me drop the cover and decided to return it to me."

Do crows actually recognize specific people? Do they know that this girl is Gabi and this woman is Lisa? Maybe they just fly into the garden for meals because they have an interest in it and they don't care about the rest? Maybe all this is a game of imagination, a biased interpretation of random events? Rather not, as scientific research shows that corvids, which include crows, do recognize human faces and silhouettes. This is confirmed by the results of the work of Korean scientists published in June 2011 in the scientific journal New Scientist.

A group of researchers led by Dr. Sang-im Lee from the National University of Seoul for several years conducted observations of magpies and found that these common - also in Poland - corvids birds can distinguish people even when they change their appearance with clothing and headgear. A series of experiments showed that the magpies sounded the alarm as soon as they noticed that a person considered dangerous was near their nests. They also flew up to the person screeching loudly, as if they wanted to drive him away. The enemies were students who, as part of the project, pretended to scare birds and try to climb trees to get to the young in the nests. The other students were indifferent to the magpies, and the third group was throwing food to the birds. And although the experimenters changed clothes and covered their heads with caps and hats, magpies flawlessly recognized people considered to be a threat to the chicks. Dr. Lee believes that magpies have learned to recognize human facial features because it helps them survive in an urban environment. And since magpies can do it, no wonder that this very useful skill has also been mastered by crows, their close relatives. Dr. John Marzluff, a professor of natural sciences at the University of Washington, who specializes in ornithology, especially in the field of corvids, has no doubts as to the existence of trade between Gabi and Lisa and the crows.

"If you want to engage with crows, you have to consistently reward them," he explained, commenting on the case of Gabi Mann in an interview with the Washington Post's website. "What is the best reward for crows? High-energy food, such as peanuts. You have to put them in the same place and at the same time of the day, making some noise to attract the bird's attention. The repetition of this procedure is extremely important because it reassures the crows of our intentions."

Marzluff, along with another American naturalist, Dr. Mark Miller, conducted research based on observing the relationships between crows and humans. It turned out that a very personal relationship can be established between the representatives of both species.

"We can certainly talk about communication taking place in both directions," Marzluff said. "Both humans and birds understand the basic signals of the other side. Thanks to the repetitive behavior of a specific person, the bird begins to trust him. We have noticed many instances where wild ravens that were regularly fed by the same people sometimes left objects where they were fed. But there is no guarantee of receiving gifts from crows - I never received such a gift. However, I have seen many items that ravens or crows have brought to other people. In these matters, you have to be a careful observer, because the bird provides man not only with shiny objects. It may well be food obtained by corvids, such as the dead chicks of other species, or a small fish or crab claws."

The case of Gabi and Lisa Mann becoming friends with crows (and especially the trade they conduct) reminds us that intelligent creatures of other species live right next to us. Raven birds, especially crows and magpies, accompany us all the time, whether in the city or in the countryside, but because they are common, we rarely pay attention to them. We explore the cosmos to find brothers in reason on distant planets, and it turns out that wise creatures live nearby, ready to enter into close and thought-provoking relationships with us.

Mammoth of the vizier Rechmire

Rechmire served as a vizier during the reign of two more pharaohs, Tuthmosis III (reigning in the years 1458-1425 BCE) and Amenhotep II (1425-1397 BCE). In ancient Egypt, the vizier was the most important government official who was answerable only to the ruler, therefore Rechmire, who held this position for decades, was an extremely powerful person. When he died, he was buried with royal splendor. The mortal remains of the dignitary were buried in a tomb in today's Sheich Abd el-Gurn in Egypt's Western Thebes.

Almost 3,400 years have passed since then. Archaeologists unearthed and examined the tomb. Even in spite of the rank of this find, it would be a topic of interest only to Egyptologists, if not for one of the paintings on the walls of the tomb.

Among the human and animal figures depicted there is a peculiar image. According to some experts, it represents a mammoth.

The Egyptologist Baruch Rosen was the first to notice this painting. In 1994, Dr. Rosen published an article in the scientific journal Nature in which he described a particular animal depicted on the wall of the tomb: a hairy elephant with long tusks, reaching people's waist. He was probably shown during a procession during which Syrian merchants (or magnates) paid homage and gifts to the pharaoh. Rosen added that in ancient Egypt both African (Loxodonta africana) and Asian elephants (Elephas maximus) were known, but the animal in Rechmire's tomb did not resemble any of these species.

Years later, Dr. Darren Naish, British science promoter and palaeozoologist from the University of Portsmouth, took up the case of the mysterious painting. In the analysis presented in January 2011, he noticed that the small elephant-like animal is covered with hair, has a characteristic bulge on the back of the head, and solid canines and these features may suggest an unexpected zoological interpretation. "There is no way it is an image of a young African or Asian elephant. The only meaningful explanation is that a stunted woolly mammoth was portrayed in Rechmire's tomb, "Naish wrote in his discussion.

However, the question is, where could the ancient Egyptians see stunted mammoths? However, if they did encounter them somewhere, it would have profound consequences for science. It would mean that the Egyptians maintained trade relations with peoples inhabiting distant Siberia (because traces of these animals were found there) and that in such a distant era a living mammoth was transported from the other end of the world.

Does the hypothesis linking Egypt with mammoths make sense? Discoveries in recent years show that the mystery can be solved.

The giant woolly mammoths (Mammuthus primigenius) became extinct at the end of the last ice age, which is 12–8 thousand years ago, as did the vast majority of the then megafauna. Smaller varieties of these archaic animals survived longer, especially the population of pygmy mammoths (Mammuthus exilis), an example of which was the 1.2 to 2.4 meters at the withers, which lived 6,000 years ago in the Channel Islands archipelago off the coast of California.

On the other hand, the last woolly mammoths, stunted as a result of living on islands or in an isolated environment, were even smaller, reaching only about 1.2 meters at the withers. One of their populations lived on the Wrangell Islands in the Siberian Arctic until 3700 years ago, and probably even several hundred years longer, that is, it survived to the time when Rechmire was the vizier.

We already know that in the heyday of ancient Egypt, mammoths still lived in northern Asia - in the form of stunted woolly mammoths. If we add to this numerous archaeological finds proving that in the 3rd and 2nd millennium BC were able to make long sea voyages, the question arises whether it can be assumed that "our" mammoth was brought from the Siberian Arctic coast along the route around Europe and the Mediterranean Sea to Egypt? This is unlikely, especially due to the extremely severe navigational conditions around the Arctic waters and the lack of archaeological traces of ancient merchants' interest in the shores of northern Asia.

But then people traveled not only by water. According to experts, land trade routes in ancient times connected distant territories - from China to Western Europe and Africa. Importantly, all roads in that era led not to Rome but to Egypt. The Egyptians knew elephants very well, but the "elephant", a meter tall, and covered with fur, could be considered a real attraction at the Pharaoh's court. It is not without significance here that Egyptologists find that the alleged mammoth is led by merchants from Syria or officials managing that land shown in the painting (during the reigns of Tuthmosis III and Amenhotep II, Syria was part of the Egyptian empire). This clue seems to confirm that the animal reached Egypt from Asia, perhaps even as far as Siberia.

However, also in the case of an overland road, it should be noted that the transport of an animal from the ice-bound Siberian North to hot Africa would be a huge commercial and logistic undertaking. So great that perhaps it is rather impossible to carry out.

However, the mammoth portrayed in the tomb of vizier Rechmire could have reached the Pharaoh's court from a place much closer. Scientists have established that on many islands in the Mediterranean Sea, such as Malta, dwarf elephants lived even several thousand years ago. The same animals survived on the Greek island of Tilos (located between Rhodes and Kos) almost until the end of the 3rd millennium BC. There were, however, significant differences between dwarf elephants and stunted woolly mammoths - for example, their hair. Naish found an explanation for this issue as well - he stressed in his interpretation of the painting that some scholars suspect that some of the extinct dwarf elephants from the Mediterranean islands were in fact stunted mammoths. The argument here is that the finds of their bones are sparse and incomplete, so it is uncertain whether the alleged dwarf elephants were covered with hair and had the skull structure characteristic of mammoths.

The mystery of the painting from Vizier Rechmire's final resting place may be solved if, as Darren Naish suggests, further research proves that the dwarf elephants from islands such as Tilos were in fact stunted mammoths. So far, the British palaeozoologist's hypothesis is only a research proposal.

Gravure with a Baltic monster

It was used to say that every sea has its own monster. From time immemorial, seafarers have told of the terrible Scylla and Charybdis lurking in the Mediterranean, the enormous sea serpent found in the Atlantic, and the tentacle-armed Kraken prowling the North Sea. Only the Baltic Sea seemed to be devoid of a "monster" legend. This situation was changed by the discovery of a forgotten illustration in the 16th-century book Les oeuvres d'Ambroise Paré (Works by Ambroise Paré).

As the title indicates, the author of the book, published in 1585, was Ambroise Paré, a French medic, and naturalist. Gravure of his work shows a giant sea snail that was supposed to exist in the waters of the Sarmatian Sea, as scientists called the Baltic Sea, citing the writings of classical authors.

The description accompanying the figure states that this is what 'a zoological anomaly about which scientists know nothing' looks like. Paré, describing the sea monster from the Baltic Sea, referred to the account of his friend, Father André de Thévet, a French Franciscan, traveler and cosmographer who lived in 1516–1590. According to de Thévet, the Baltic snail depicted in the engraving was the size of a wine barrel. He wore an enormous shell on his back and horns the size of a deer antler on his head. The creature's mouth was shaped like a cat's, and the beast's eyes were not at the end of the pistils, as is customary for snails, but on the sides of the head. A French priest described another unusual feature of the giant snail to the doctor: this animal, unlike the typical snail, had four legs ending in fingers or protrusions.

The unusual creature was supposed to spend most of its time at sea, but every now and then it would come ashore to eat land plants. Paré, as befits a medic, became interested in whether the snail could be used in healing, and then, based on de Thévet's account, wrote that the blood of a mysterious creature soothes the suffering of leprosy patients. And since Paré was a real French, he also drew attention to the culinary qualities of the Baltic monster - apparently, the meat of the giant snail was considered very delicate and tasty.

There is only so much information about the terrible snail in the book, so it is worth taking a closer look at the author of the source.

Can Ambroise Paré be considered a credible researcher? He lived in the years 1509–1590 when the foundations of modern science were just emerging. Paré was not one of the many compilers of ancient (and largely outdated) knowledge at that time, but he was a real researcher who left behind dozens of scientific works - medical and natural. He revolutionized the methods used in surgery, for which he was appointed the royal surgeon, who was under the rule of four successive rulers of France (Henry II, Francis II, Charles IX, and Henry III), i.e. for almost the entire second half of the 16th century. Today he is considered one of the fathers of modern surgery.

It, therefore, seems that the author of the source about the giant snail is a credible researcher. The problem is that he did not know this creature from an autopsy, and relied on the stories of a

Franciscan friend. That André de Thévet also seems to be a credible informant, as he is valued for descriptions of distant lands he visited (the Middle East, Palestine, Greece, Egypt, Turkey, and Brazil). However, as the biography of this scholarly traveler shows, he did not travel to the Baltic Sea. As a result, he could not be an eyewitness to the "wonders" of the Sarmatian Sea, that is, he must have heard the story of a snail from a sailor who boasted that he had seen a monster at the northern end of Europe. And we have no news about this informant, so from the scientific point of view, the message about the giant snail does not deserve faith, at best it can be considered a marine legend. But there is also a problem with this because about the snail with horns not a single story has survived by those sailors who traveled the Baltic every day - you can not hear anything about it in the folklore of German, Swedish or Danish sea wolves.

The image of the beast was recorded only in a French source, i.e. written in a country whose ships rarely ventured into the Baltic Sea. So where and when could de Thévet hear the rumors of a Sarmatian Sea snail? It probably happened aboard one of the ships of the small fleet of French Admiral Nicolas Durand de Villegaignon, who sailed in 1555 to conquer Brazil. André de Thévet participated in this expedition and during it, he could hear many extraordinary sea stories.

Let us assume, however, that de Thévet's message is interesting enough to be dealt with also in terms of mythology, and then cryptozoology.

The creature Paré describes has a shell that is characteristic of molluscs, but all the rest of its features are typical of herbivorous vertebrates - a puzzling combination. Modern science does not know such an animal, so perhaps we are dealing with a fantastic creature, in other words: belonging to the demonology of the Baltic peoples originating from paganism?

Probably not, because neither in Scandinavian, Baltic, and Slavic demonology there is a terrible snail with horns. Besides, Paré did not give any of its supernatural properties, i.e. there is no basis for considering the described creature as a realistic vision of a northern European water demon from folk beliefs.

Karl Shuker, the author of the best-selling books describing the mysteries of cryptozoology, tried to unravel the mystery. In a note published on April 16, 2012, on websites devoted to this type of issue, Shuker hypothesized that the giant Sarmatian Sea snail is a species unknown to science of shellless nudibranchs. Only such a taxonomic classification would, according to Shuker, explain the unusual number of beast legs for mollusks. What's more, floating nudibranchs do sometimes take fantastic shapes, in which you can see paws, horns, and other non-snail anatomical details. The weakness of Shuker's hypothesis is that the floating nudibranchs are much smaller than the monster described by the French medic, and they have no shells, so the differences are fundamental. The largest of the nudibranchs that traverse the sea, usually measure just over half a meter, and the longest of them grows up to 120 centimeters at most. Moreover, such large specimens are found in warm, not cold waters. Currently, only a dozen or so species of molluscs live in the Baltic Sea and all of them belong to the "general cargo".

On the other hand, among the shelled snails, Syrinx aruanus from the waters of northern Australia, eastern Indonesia, and Papua New Guinea is considered the largest. The shell of this mollusk is nearly a meter long, and the syrinx itself weighs almost 20 kilograms. Although it is a large creature, it is far from the beast described by Paré, and otherwise, it can only survive in tropical waters, so there is no way that it will appear in the Baltic Sea.

It follows that there are no traces that would allow us to identify the Baltic monster either with folk demons or with real animals. After all, let's list the data that Paré has given us once again.

According to him, the mysterious animal was the size of a wine barrel, and since in 16th-century France typical barrels had 225–228 liters of capacity, it can be assumed that the horned snail weighed no less than a quarter of a ton. Besides, it had four paws, a cat's face, eyes situated like those of vertebrates, and on its head resembling horns. All these features point to a water reptile. Typically the snail shell on the back of the animal is troublesome. Maybe it was not a shell, but an armor or a dorsal shell, which the author of the engraving could not properly imagine, so he showed it like as an escargot, well known from French tables?

However, even taking such an assumption into account, we would be dealing with a powerful and dangerous aquatic reptile, and such do not occur in our region of the world. It is also difficult to say that it could be a real species that once existed, but unknown to science, because large reptiles adapt very badly to the climatic conditions of the Baltic Sea - this is not their geographical zone.

Unless we assumed that a wealthy merchant returning from Egypt brought a tiny crocodile as a souvenir, and when it grew older, the terrified owner threw it into the sea.

Such a monster could indeed be seen by a few people, but it would only last for a few months as it would die in the first cold winter.

Although we have analyzed the matter from every possible angle, we still know about the mysterious creature only as much as can be seen in the 16th-century gravure and as indicated by Paré's general description. So are we dealing with a limited-scale seafaring legend or a monster from forgotten demonological tales? It is only certain that the French scientist who described the huge Baltic snail in his book was convinced of its real existence. That is why he left a visual message in the form of a mysterious artifact - a gravure depicting a mysterious creature from the Sarmatian Sea.

How far from legend to facts? From speculation to hard evidence? There are artifacts the existence of which science confirms, but has no idea how to interpret them, so artists deal with them, and those used to let their imaginations run wild. Extraordinary artifacts inspire and provoke writers, artists, and filmmakers who sometimes interpret them freely.

In turn, scientists are trying to unravel the mysteries concerning strange finds and separate the originals from the falsifications.

Map of the Lost World

The map, carved on the surface of some round stone, shows the outline of the land, mountain ranges, and river beds. Some points were also marked, perhaps settlements or sanctuaries. Unfortunately, the topographic details do not fit into any map developed by modern cartographers. What land is represented by the mysterious map carved on one of the famous stones from Ica? Maybe the legendary - and thus inspiring writers and filmmakers - Lost World?

Ica is a large Peruvian city located about 300 kilometers south of Lima. In the center of Ica, in a house in Plaza de Armas, lived Dr. Janvier Cabrera (died in 2001), who had amassed a collection of thousands of engraved stones. The smallest are several centimeters in diameter, and the largest cannot be lifted by one person. Drawings of unusual content have been carved on the surface of these pebbles. It was this content that made scientists hesitate to research artifacts from Ica, which they unequivocally consider to be forgeries. Maps of unknown lands are carved on several stones. Others have been engraved with people in ancient Indian clothes who observe the sky with a telescope and examine small objects with a magnifying glass. Surgical operations, including heart and kidney transplants, as well as blood transfusions, can be diagnosed. The drawings of animals are even more surprising. Against the background of vegetation from ancient times, there are drawings of mastodons, huge birds, kangaroos, and camels. There are also reptiles of the Mesozoic era: Stegosaurus and Brontosaurus. The strangest thing is that animals are accompanied by humans. In one of the drawings, hunters hunt a giant reptile using pointed tools, and the ancestor of today's birds, Archeopteryx, hovers in the air above their heads.

Pictures of the Ica stones have been circulated around the world thanks to Robert Charroux, who in 1974 publicized the matter of the finds in his book The Secret of the Andes. In the following years, Erich von Däniken also wrote about the stones from Ica. Today it is known that some of the rites are certainly modern fakes, but a group of local teachers took the risk of being ridiculed and carried out an expert opinion on some of the stones at the Higher Technical School in Lima. It turned out that some of the engravings are very old - several thousand years old. Among them were also those depicting scenes involving humans and dinosaurs. The problem is that the results of the analyses have not been verified in other research centers, and this is the only way to get arguments for discussions with archaeologists or paleontologists.

Another trace of the Lost World and a civilization that may have flourished on the lost land outlined in one of the maps of Ica, controversial in the light of science, was found in Acambaro. In 1945, while traveling through central Mexico, the merchant Waldemar Julsrud came across an underground vault in Acambaro, which had been unveiled during heavy rains. There was unusual earthenware in the hiding place. Over the next seven years, Julsrud was to retrieve as many as 33,000 10 to 15 centimeters high figures from subsequent underground storage spaces.

They are made of clay fired at 500 degrees. They depict people and animals, and even whole groups of them, and also in this case we are dealing with creatures resembling plesiosaurs, stegosaurs, or brontosaurs. However, the Mexican finds do not show the hunting of giant reptiles. The alleged dinosaurs appear domesticated as they are cuddled, fed, and ridden by children. It is difficult to remain indifferent to such representations, which is why for some they are a document of an episode of human history unknown to science and for others a brazen forgery.

Julsrud for many years tried to interest his collection of archaeologists, but - like the stones from Ica - clay figurines were considered modern counterfeits. It was only in the 1950s that the American amateur historian Charles Hapgood took up the matter, devoting the rest of his life to researching the Julsruda collection. Apparently, he also found 44 clay figures - in a ditch under the house of the local police chief. When he analyzed the places of the finds, he realized that most of the artifacts were discovered in an area of only half a hectare, where there were as many as several hundred separate hiding places. With this in mind, some ancient mystery enthusiasts say that, as with Ica, it was a sort of archive left to posterity.

Hapgood also discovered that almost identical figures have been excavated for years by Indians in nearby San Miguel Allende, but not directly from the ground, but from inside the ruins of some ancient structures. Hapgood claimed to have commissioned research on the finds. Reportedly, analyzes using the C14 carbon isotope determined the age of the figures at 6.5 thousand years, and dating using the thermoluminescence method - at 4 to 4.5 thousand years.

Hapgood, therefore, believed in the authenticity of the Acambaro statues and claimed that scientists ignored these monuments because they did not fit the official picture of human history.

The third site where equally mysterious representations have been found is the Caverna da Pedra Pintada (Painted Stone Cave) cave archaeological site on the Parima River in northern Brazil. Among the many symbols fixed on the rock, such as crosses, snakes, suns, animals, people (dated by the thermoluminescence method to about 10,000 years ago), reportedly - according to enthusiasts - there are also drawings that can be interpreted as stylized representations of Mesozoic reptiles.

Other alleged depictions of dinosaurs were found in Utah in the 1990s while identifying caves where ancient Indians painted and carved rock paintings. Of particular interest is the iconography of the Kachina Bridge grotto in the Natural Bridges natural monument formation. The carvings turned out to be so surprising that their interpretation and dating led to disputes in the scientific community. And what do the paintings from the Kachina Bridge show? Some researchers say that the reptiles resemble dinosaurs, others that they are the s-shaped outlines of sacred river backwaters...

In March 2011, the dispute over the interpretation and dating of this find flared up again. For many cryptozoology enthusiasts, however, there is no doubt that the Kachina Bridge petroglyphs represent diplodocus-like herbivorous dinosaurs. One of the advocates of this interpretation of the paintings is Dr. Phil Senter, a lecturer in biology at North Carolina State University in Fayetteville.

"The first time I saw the paintings, I had no doubts," Senter told Discovery News. - The animals shown there look like sauropods.

Do the cave paintings actually represent dinosaurs living at the same time as humans? And if they are not dinosaurs, what creatures have the Indian artists immortalized on the rock wall?

This is not the end, however, because the images of animals that looked like dinosaurs were painted and carved not only in America.

Two of these finds are relatively young and easy to date. Ta Prohm Temple is located in the jungle next to Angor Thom, the last capital of the medieval Cambodian empire. Ta Prohm was built about 800 years ago.

Its entrance is decorated with bas-reliefs with motifs characteristic of both Hindu mythology and Buddhism.

Among them is a depiction of an animal with a special appearance - a tough-bodied quadruped with a row of plates on its back. Only one animal known to science looked like this only that it died out 65 million years ago.

Also known in Poland for his books on cryptozoology, Dr. Karl Shuker claims that he discovered this bas-relief in the fall of 2007.

According to this researcher, a petroglyph from the temple's pediment depicts a Stegosaurus, a dinosaur known to paleontologists from many fossils. But how did a Cambodian sculptor from eight centuries ago know what the Stegosaurus looked like and why did he place its image in a holy place? "Does the Ta Prohm relief prove that some species of dinosaurs survived 65 million years ago and still lived 800 years ago in the Cambodian jungle?" Shuker asked the embarrassing question.

The fact is that the image of this animal resembles the reconstructions of the stegosaurs made by paleontologists.

Did the descendants of ancient reptiles considered extinct recently lived in the wild corners of Cambodia? What if the artist immortalized another animal in stone? But what animal?

Similar problems are caused by the painting from the 16th century, the analysis of which in the fall of 2011 was published on many cryptozoological websites. The painting depicts the temptation of Saint Anthony and is in the collection of the Royal Museum of Fine Arts in Antwerp. The painting is the work of Marten de Vos, a Flemish artist who lived in the years 1532–1603.

According to a religious legend not entirely consistent with the facts, Saint Anthony the Great went to the desert to pray in solitude. In the outback, he was tempted by a legion of demons of all shapes. Many artists reached for this topic, and in the 16th century, demons were usually depicted as bizarre chimeras. However, de Vos's painting has a special distinguishing feature: at the bottom of the painting, on the right, there is an animal that looks like a dinosaur.

It is worth paying attention to the composition of the picture. All the demonic creatures swirl above and below the painting.

At the bottom, in the foreground, real animals are visible. So there is a pair of lions, there is a pig. In the same group, the painter placed a creature with the appearance of a dinosaur.

Are cryptozoologists right to suspect that Marten de Vos immortalized the great reptile in his version of the temptation of Saint Anthony? But where could the artist see such an animal? Or maybe he painted them according to the description of a witness who met the beast during an exotic journey? Maybe a trip to the Lost World?

Artifacts depicting large reptiles are surprising, but do they actually prove the existence of enclaves where such living fossils have survived? Maybe the images - assuming that they are all monuments and not fakes - are only a figment of their creators' imaginations, or we, associating these images with our knowledge of extinct reptiles, misinterpret them? Because until recently there was a Lost World somewhere, where is the evidence, for example, the skeletons of dinosaurs that lived thousands or even hundreds of years ago? After all, you do not need to find living individuals, you only need the remains of great reptiles dating back to the time when our species already existed.

And it was precisely such a discovery that was troublesome for science that the website "Ancient Origins" announced on January 10, 2015.

According to this source, in May 2012, a fossilized horn of a triceratops was found in Dawson County, Montana, and found its way to the Glendive Dinosaur and Fossil Museum (also Montana). It was only two years later that Dr. Hugh Miller, head of the Paleochronology Department, began investigating the find.

Two samples of the triceratops horn were sent to a Georgia State University laboratory, where radiocarbon analyzes revealed an unusual age for the find: the dinosaur, the remains of which were examined, lived between 41,000 and 33,000 years ago.

Miller stated that the surprising dating of dinosaur remains is not unique, as many paleochronological analyzes yield similar results, which are, however, rejected as being treated as anomalies or errors. The paleontologist from Glendive emphasized that until recently the C14 method was not used to study dinosaur remains, because it is effective when the biological material is no more than 55,000 years old.

According to this expert, serious methodological flaws lead to situations such as the 68 million-year-old discovery of soft tissue in the Tyrannosaurus leg bone presented in the March 2005 journal Science. Miller assured that this research success announced by paleontologist Mary Schweitzer and her team was due to a measurement error, as there is no way that soft tissue will not become fossilized in more than a million years. In his opinion, the dominant group of paleontologists blocks the presentation and publication of research materials showing that some dinosaur remains date back only several dozen millennia.

Knowing about the doubts that scientists themselves have, let us consider how to explain the creation of the artifacts described above since there are analyzes dating back to tens of thousands of years of dinosaur remains? There are three hypotheses at play - however, it should be remembered that they were all constructed not by scientists, but by enthusiasts of the mysteries of the past.

The first hypothesis is that human history goes back many millions of years and that rational man already existed in the era of the great reptiles. The second is that some dinosaurs survived the global catastrophe 65 million years ago and lived in enclaves until almost modern times. The third is that the scientifically debatable finds are the legacy of some lost civilization that flourished on an equally lost land.

Is this the unidentified area of the third hypothesis, the Lost World, indicated by the map carved on the Ica stone? If we assume that it is authentic, let's try to confront it with theories about lost lands and civilization.

Let's start with the most famous place in mass culture - thanks to Arthur Conan Doyle's novel - that is associated with a "reserve" where dinosaurs survived.

It is, of course, the place described in the novel "Lost World" from 1912, which saw many reprints and several film adaptations. The suggestive description of a mysterious enclave of fauna and flora presented in it from millions of years ago captured the imaginations of mystery researchers, travelers, and naturalists. Many of them made the effort to find the most real Lost World.

They did not follow blindly but were guided by rumors circulating since the nineteenth century about unusual fauna found in one of the desolate places of South America. This mysterious place was to be located in Venezuela on the savannah Gran Sabana plateau, where imposing tepui ("houses of the gods" in the language of the local Indians), ie specifically shaped rocky plateaus. Mystery Hunters point to two of them as the source of the legend of the Lost World - Roraima and Auyan-tepui. I will only deal with the latter here, because quite a lot has been written about Roraim, also in Poland.

Auyan-tepui is a vast (700 square kilometers) table mountain covered with a jungle at the top, from which the Salto Angel, the highest waterfall in the world (almost a thousand meters high) flows down. One of the people who said that they saw unusual animals on Auyan-tepui was the late Alexander Laime, a Latvian who lived for many years at the foot of the plateau. It is said that he saw miniature plesiosaurs twice during his trips. Also, Professor Fabian Michelangeli, a biologist and physician at the Venezuelan Institute of Scientific Research, claimed that in 1998 he personally arrived at the plateau where he saw the same creatures.

The alleged Venezuelan Lost World is actually a very hard-to-reach area - to get to it, you have to climb vertical rock walls 1.5 to 2.5 kilometers high. As a result, this plateau is very little known - for example, the magnificent Salto Angel waterfall was only discovered in 1933. Even today, only experienced climbers can climb the mountain. Attempts have been made to get there by helicopter many times, but the results were miserable because the surface of the tepui is so shaped and densely covered with tropical forest that the helicopter has problems landing. Besides, there are no water sources there, so it has to be transported by air.

An example of an unsuccessful trip is the professionally organized expedition of the German ZDF television team, which in 1995 went to Auyan-tepui to make a documentary about the flora and fauna of this region. The filmmakers stayed in the jungle for only a few days, as they could not count on regular supplies, including drinking water.

Everything was delivered by helicopter, which was too low for the lifting capacity and proved to be technically unreliable. The Germans did not film any mysterious animals, although during one of their helicopter flights they allegedly saw in distant lake reptiles resembling plesiosaurs, but small in size.

Auyan-tepui is not the only place where rumors place the Lost World, so maybe the map from Ica shows a completely different area? For example, Foggy Island in the fog about to lie in the Indian Ocean halfway between Africa and Sumatra?

I mention Skull Island not by accident, because, like the heights from Conan Doyle's Lost World, it has grown into mass imagination, thanks to the films and the novel King Kong by Delos Wheeler Lovelace.

Lovelace, an American journalist, and writer who lived in the years 1894–1967 wrote and published an amendment to the film script in 1932 before the cinematic film about the great monkey hit the screens (King Kong, 1933, directed by Merian Cooper and Ernest Schoedsack). There are many threads typical of adventure literature in this story, but the background of the plot is Skull Island, a tropical land lost in the ocean. This is another Lost World, where apart from a huge and quite intelligent monkey, there live other creatures of great size, including dinosaurs. The story, like other stories about the lost worlds, enjoys an unflagging interest in subsequent generations and returns to the screens from time to time in subsequent adaptations. King Kong usually looks like a giant ape but moves like a human when upright. Depending on the creativity of the filmmakers, this creature was from a few to even 30 meters tall. The smallest King Kong in the history of cinema is the hero of the production from 2005, directed by Peter Jackson - he is "only" 24 feet tall, or about 7.2 meters, and looks and moves like a gorilla. At the same time, it is closest to the monkey in Lovelace's novels and the creature that actually once lived on Earth. Can the existence of a place like Skull Island and a humanoid being a few meters high be taken seriously? What do scientists think?

In an interview with Reuters on January 2, 2006, Dr. Sue Liebermann, evolutionary biologist and head of the WWF's global species protection program, said that such giant creatures may have lived on uninhabited islands until relatively recently. In many such isolated places, there are cases of extreme mutations, both animal dwarfism, and gigantism, i.e. excessive body growth. Serious changes can affect both individual animals and whole species.

An example is the Komodo island dragons. The largest lizards in the world are over 5 meters long. On the other hand, on the island of Gough in the South Atlantic, there are giant mice, and in Madagascar several hundred years ago there were flightless birds reaching over 3 meters and lemurs weighing over 80 kilograms.

But what about the already mentioned prototype of King Kong? Science has established that giant anthropoids did indeed live on Earth. These creatures were discovered in 1935 by the Dutch paleontologist Gustav von Koenigswald and called the gigantopithecus, and he defined the area of occurrence as China and Southeast Asia (later it turned out that related varieties also lived much further in the west, through India to the Mediterranean region). The oldest fossils of giant humanoid creatures are 8 million years old, and the youngest is 100 millennia old, with some scientists admitting that the dating of these finds is doubtful and they may well be much younger, even several thousand years ago. Importantly, scientists still do not have a complete skeleton for any of the Gigantopithecus varieties.

Initially, researchers included these creatures in the human lineage, but over time it was recognized that they were an ape. The youngest variety of gigantopithecus, Gigantophitecus blacki, resembled both an orangutan and a gorilla with its body proportions and structure, but it was also able to move upright. The known remains of this creature date from between a million and 100 thousand years ago.

This gigantopithecus was about 3 meters tall and weighed over 500 kilograms. For comparison: males of the largest modern apes, mountain gorillas (Gorilla beringei beringei), which live in the

Virunga Mountains of Uganda, when they stand upright, reach a height of 190 centimeters. And they weigh about a quarter of a ton.

Gigantophitecus blacki lived in the upland tropical forests of Asia and ate plants, mainly bamboo, and hard-shelled fruit. Some researchers have suggested that such a diet of Gigantopithecus may have been the cause of the growth of his powerful jaws. An important observation is also the fact that he lived at a time when a man appeared in Asia - certainly Homo erectus, and perhaps also later species. Some scientists assume that Gigantopithecus and man somehow coexisted by interacting with each other, which led to the extinction of the greatest anthropoids in history. However, the memory of the great anthropoids survives in tales of giants handed down from generation to generation.

Is it possible to build on this basis the presumption that the messages of various peoples of the tropical zone made such a great impression on European travelers that they, in turn, began to tell each other about great monkeys and hence the Hollywood King Kong was created? The paths of development of myths and legends are so complicated that it is possible to imagine such a development pattern of this story.

Another literary Świat Zaginiony was created by Edgar Rice Burroughs in his 1918 novel Land Forgotten by Time, which he quickly developed into a trilogy (the film was adapted in 1977, 1979, and 2009).

The story takes place during the First World War. The survivors of the submarine land on Caspak Island, also known as Caprona. On this tropical jungle land somewhere in the South Atlantic (near Antarctica), dinosaurs also live.

Literary lost worlds - such as Conan Doyle's Plateau, Skull Island, and Caspak Burroughs - need not be fiction. Such enclaves may and may still exist, such as somewhere in the Pacific, where there are still plenty of deserted and virtually unexplored islands. Or maybe the method of searching for the Lost World is not correct? Maybe instead of moving across lands and oceans, we should go to the center of the Earth? As several visionaries have claimed - not only writers but also people who consider themselves philosophers and creators of cosmogonic theories - there are great empty spaces in the depths of the Earth, where mysterious life forms flourish and mysterious civilizations hide.

This concept has long fascinated writers, researchers, and adventurers. Has the Frenchman Jules Verne indicated the gate to the underground land? Or perhaps the German Commander Brodd and the American Admiral Byrd discovered it and entered the Lost World through it? Is this surprising direction indicated by the map from Ica?

The idea of the inner worlds was most popularized by Jules Verne's novel "Wyprawa to the Interior of the Earth", which was first reprinted in 1864 and was screened many times. The French writer placed the entrance to the underworld in the crater of an extinct volcano beneath the Snaefellsjökull glacier in Iceland. Through the crater of another volcano, Stromboli in the Aeolian Islands in the Mediterranean, the heroes of the novel left - safe and sound - the underworld.

Importantly, the idea of this story did not arise from anything, because the concept of an empty Earth, i.e. the existence of some land within our globe, was presented long before Verne was born. The hypothesis that the Earth is empty inside was already taken up by Plato. In the 17th century, the famous astronomer Edmund Halley was inclined to this view, proposing the assumption that our globe is made of as many as four spheres.

This topic was taken up in 1818 by John Cleves Symmes, a retired infantry captain. Symmes gave the US Congress an official letter saying that the Earth is empty inside and that the Americans should colonize the territory. The retired officer intended personally to lead the first expedition. He demanded equipment from the authorities, hundreds of people, reindeer, and elephants. So equipped, he planned to break through the gate to the underground Lost World, which, in his opinion, was in Siberia. American politicians rejected this request.

In 1906, another hollow-Earth promoter, William Reed, hypothesized that our planet's crust was 800 miles thick. Below this layer is an empty space 6400 miles in diameter, illuminated by a core that acts as the inner sun. According to Reed, however, it was not necessary to drill through 1,300 kilometers of rock, because there were two entrances to the interior of the planet, one at each of the Earth's poles.

The esoteric tradition also mentions land in the underground sphere of the globe. It is Shambhala, an enclave of sages who - depending on the interpretation of the mystical school - are either immortal or form a tribe of the elect highly developed spiritually. The people of Shambhala were to live in hiding for millennia and look after (or just watch) humanity.

Only the initiated could get into this land, and a secret gate supposedly existed somewhere in Asia.

The Reed concept and the Shambhala tradition were combined in the interwar period with truly remarkable consequences. In 1931, Professor Herman Wirth, a German researcher dealing with the issues of racial purity, proposed a fateful hypothesis. According to him, in the distant past, a giant piece of cosmic ice (called by Wirth the Ice Moon) fell on the Aryan-inhabited Atlantis and destroyed it. Soon the Nazi dark esotericists developed this theory in their own way.

In 1935, Wirth met with Reichsführer Heinrich Himmler, who fell in love with the professor's concept.

Soon it was supported by the empty-Earth hypothesis and a kind of variant of the Shambhala legend. In the circles of supporters of German National Socialism, a mystical-intellectual movement arose which proclaimed that the original Aryan race, that is - according to the Nazi ideology - the progenitors of the Germans, was born in Atlantis located north of Europe, possibly near the North Pole. After the destruction of Atlantis, its inhabitants moved to an enclave inside the Earth and are still hidden there. The concept of an empty Earth and the pro-German civilization hidden deep beneath its surface fascinated not only Himmler, but also Adolf Hitler himself. Convincing such prominent people about the truth of this assumption resulted in an extraordinary mission.

According to some mystery researchers, in May 1943 a submarine was sent from Germany towards the North Pole, where there was to be a hole leading to the underground land. The U-209 was reportedly commanded by Captain Heinrich Brodda, who had ideologically certain sailors and scientists on board. There was a war going on, there was heavy fighting in the Atlantic, so Broddy's mission was doubly risky.

Nothing else is known about the fate of the U-209 and its crew. One thing is for sure: the u-boot has not returned from the expedition.

However, this is not the end of the story about the Lost World in the depths of the earth. Information about the U-209 expedition has been preserved in the archives and may have inspired the people who led to another expedition. Whether it was so, it is not certain, but the fact is that three and a half years after Brodd, for a similar purpose, but in the opposite direction, a great American military expedition set off. Secret expedition and equipped for war.

The expedition of American ships to Antarctica was organized by Rear Admiral Richard Evelyn Byrd, born in 1888 and died in 1957. He had long been interested in ice-covered regions beyond the reach of civilization, for example, on May 9, 1926, he flew over the North Pole by plane, which was then a pioneering feat.

It is not known how Byrd convinced the general staff and politicians that the US would pay the enormous costs of a military expedition to the pole, this time to the south. Presumably, this decision was influenced by various secret reports, including documents from Nazi archives. Perhaps they also recalled the offer that John Cleves Symmes had made to the US Congress 150 years earlier. The official name of the expedition was: The US Navy's Antarctic Development Program, but its guidelines also included a project to carry out a super-secret action called "Operation Highjump" ("Operation Highjump").

The convoy left the United States on August 26, 1946, and the entire operation lasted until March 1947. Admiral Byrd's expedition consisted of 4,700 men, 13 ships, and a dozen aircraft. In fact, it was as if the admiral flotilla was going to war. Officially, the expedition was to conduct personnel and equipment exercises in the Arctic Zone and to prepare analyzes of the possibility of building and maintaining an American base in Antarctica. However, the secret assumptions assumed that the members of the expedition were to check whether there was a hidden Nazi military base in this area, called Station 211 by researchers of secret operations from World War II. Apparently, the search for Station

211 was carried out from the sea, land, and air, but this base Byrd didn't discover. Instead, the Americans stumbled upon a special place, which has passed into the circle of historical mysteries under the name of Bunger's Oasis. Byrd, in his diary, fragments of which became known to the media after the admiral's death, was to describe an unusual land in the South Pole region seen with his own eyes. During one of his ferry flights in February 1947, Byrd reportedly saw "great patches of color" emerge from behind a range of ice-covered mountains. This is how the area later called Bunger's Oasis was found, after the name of the first person to spot it - one of the pilots participating in the expedition, Lieutenant David E. Bunger.

According to the admiral's records, the "color spots" turned out to be a vast valley devoid of everlasting ice. Byrd saw lakes, lush vegetation, and giant animals. On February 19, 1947, according to not entirely reliable sources, the commander of the expedition noted the following in his diary: "At 10:00 we flew over the mountain range. Behind him, the landscape changed. You can see the valley with the river running through its middle part. The valley is green with vegetation. Magnificent forests grow on the slopes of the mountains. Our navigation instruments give inaccurate readings because we are close to the Pole. "

The next note, made at 10:05 AM, said: "We're making a sharp left turn to better explore the valley. It is green with moss or thick grasses. In the distance, I see a large animal resembling an elephant. We're reducing the altitude to 1,000 feet. I take my binoculars to better examine the animal. I confirm. The animal I see is undoubtedly a mammoth! I am trying to contact the base, but the radio is not working. "

Some mystery researchers say that it was in the Oasis of Bunger that the admiral found the entrance to the Lost World inside the Earth. And that also here during the war the U-209 of Captain Broddy set off, and the course of his ship recorded in German archives to the North Pole was only intended to camouflage the real destination of the voyage. Even if Byrd discovered the underground passage, we have no idea what the US military command and government did with such remarkable knowledge. The conspiracy theory for the secret of the admiral's expedition is losing momentum at this point as more data is lacking to allow further speculation to develop.

The situation is obscured by the fact that today it is difficult to determine how much of the known information comes from the authentic records of Byrd, and how much is made up of plotters. However, lovers of old secrets prove that Bunger's Oasis really exists, as confirmed by images taken by the ESSA-7 satellite, which was revealed in 1970. This photographic footage shows a gigantic depression at the Pole - where enthusiasts say this is the mild-climate valley described by Byrd and the alleged entrance to the interior of the planet.

Does the Ica map show any of the Lost Worlds I mentioned? Or some other because there are still many unknown places on Earth? Or maybe this whole journey down the trail of hypotheses and uncertain reports is only picturesque speculation closer to fiction in the spirit of Conan Doyle than to scientific theories based on facts? It is true that most researchers consider the map of Ica a fake, but this object was used in this article as a symbol of unsolved mysteries of the past.

Amazon figurine

The ancient Greeks knew the myth of the Amazons. It was repeated until modern times, and as the world explored, the land of brave women was moved further and further, until it finally ended up in the jungles of South America. Perhaps, however, we are dealing with something more than just colorful stories, since the discoveries of recent years prove that there are facts behind the myth about the Amazons?

Let's start with the mystery related to the Amazons, which is hidden in a bronze statue made two thousand years ago in the Roman Empire. It is not known exactly when and where the figurine was made, nor is it known who and where found it. It was "discovered" again in the 21st century among the dusty and indescribable exhibits at the warehouse of the Museum of Arts and Crafts in Hamburg. According to the description presented in April 2012 on the Live Science website, the statuette shows a slim woman, wearing only a loincloth, with long hair arranged in an elaborate hairstyle. The character is shown in motion, but it is not a dancer. The woman raises her armed arm into a curved knife in a triumphant gesture. Dr. Alfonso Manas from the University of

Granada, who examined the figurine, believes that it represents a gladiator, a fighter professionally fighting in the arena. This opinion of Manas was influenced by the fact that the woman holds in her hand a characteristic weapon called sica - a short, curved sword used first by the tribes of the Balkans, and then by gladiators. The very gesture made by the warrior is also characteristic because according to Roman chroniclers, it was performed towards the audience by the gladiator who won the fight. The Spanish scholar believes that the statue is a portrait of a specific person, probably a particularly popular warrior of the empire arenas because the face, silhouette, and attire were rendered with great precision. Originally, at the bottom of the figurine, there could have been an inscription with the person's name, but the sculpture is incomplete.

Experts assume that so few representations of women fighting in amphitheaters have survived because there were not many gladiators themselves. Apart from the Hamburg figurine, we know only the relief from the Greek city of Halicarnassus (stored in the British Museum in London), showing the fight of two armed women. The written sources about them are also sparse - only about a dozen Roman notes mention gladiatorial. One of them states that women's fights in the arena were forbidden by Emperor Septimius Severus in 200 CE.

Dr. Manas suggests that the "amazons" appearing in the arena were a special attraction for the male audience not only because there were few of them in the entire empire.

"At that time, Rome was constantly waging wars and bloodshed was something banal. However, in society as militarized as the Roman one, the use of weapons belonged exclusively to men, explained the Spanish researcher. "On the other hand, a society lady, or even an ordinary townswoman, did not appear naked in public places. Only slaves and free persons of the lowest social status could parade in this way. Therefore, naked and armed women struggling with opponents in the amphitheater added notes of eroticism to the gladiatorial shows, stimulating the imagination and libido of the viewers."

In this context, it is worth mentioning another discovery that takes us back several hundred years in relation to Roman gladiators, and besides, it seems to be closer than they are related to the issue of mythical (or perhaps historical?) Amazons.

Dr. Adrienne Mayor, archaeologists at the J. Paul Getty Museum in Los Angeles, and Dr. John Colarusso, a linguist at McMaster University in Hamilton, Canada, managed to read the names of the Amazons immortalized on Ancient Greek pottery. As reported on September 23, 2014, by "National Geographic News", a pair of scientists selected twelve vases and goblets (from the 6th and 5th centuries BCE) decorated with scenes of fighting or hunting in which the Amazons took part. Then the inscriptions on them were analyzed.

It is important to note that Colarusso is an expert not only on the Ancient Greek language but above all on the ancient languages of the Caucasus region. When he decoded the captions commenting on the scenes depicted on the dishes, it turned out that they were written in the Greek alphabet, but conveyed proper names and words from the Scythian language. In antiquity, the Scythians inhabited the steppes on the northern and eastern shores of the Black Sea, they also inhabited part of the Caucasian region.

Moreover, ancient authors located the headquarters of the Amazons precisely near the lands of the Scythians. What were the names of the warrior women of ancient Greek pottery? One was called Worthy Armor, another Dog Tracker another - Hot Thighs.

Experts assume that it could have been names as well as nicknames. According to Colarusso and Mayor, the former suited a particularly formidable female warrior, the latter suited a hunter, and the third had clear erotic connotations. A couple of scholars emphasize that neither of these names should surprise us because the Amazons seemed exotic and exciting to the Greeks because they lived in a completely different way than the women of Athens or Sparta.

If we recall the above-mentioned statement by Dr. Manas about gladiators, we will understand that the Romans took over from the Hellenes the way of looking at warlike women.

At the same time, in the case of the inhabitants of the empire, it was no longer about the "right" Amazons, but about women who resembled them in a way, i.e. champions of bloody competitions in the arena.

And another interesting discovery that may explain where the gladiator came from, who, according to Manas, was faithfully portrayed in bronze. In January 2005, the British press reported the discovery of the remains of the Amazons from the time of the Roman Empire. It was not a fresh find, as it had been in the museum warehouse for 40 years. In the 1960s, near Brougham in Cumbria (North West England), archaeologists unearthed burials, the relics of which were not examined until 2004 by an independent British team of members of the private Nottingham Barbican Research Association, led by Dr. Hilary Cool.

Artifacts and human remains came from burials found on the site of a former Roman military camp from the 3rd century CE. Archaeologists exploring this place in the 1960s assumed that they were dealing with legionnaires' graves, only Dr. Cool noticed the remains of two cremated bodies buried with horses, which were also burned. Analysis of the remains of the alleged Roman legionnaires revealed that they were women. They were between 20 and 40 years old at the time of their death. In their graves, there were pottery vessels, silver bowls, sword, scabbards pieces of ivory, and bone cladding of decorative boxes.

It was rich posthumous equipment and testified to the high social status of both buried women, as well as their affiliation to the imperial army.

"If we found the remains of the men, no one would have any doubts that they were cavalrymen serving in the Roman army," Hilary Cool said publicly. "So why should we have doubts just because it is about women?" Everything indicates that the dead belonged to the Roman cavalry, and if so, you might as well assume that they were Amazons.

Scientists speculated that the women came from the Black Sea or Danube steppes. Probably irregular units formed by warriors from the territories subordinate to Rome were stationed in the military camp under study.

Many traces indicate that soldiers from the Danube Delta area had come to Brougham, which was one of the areas where the ancient Greeks located the Amazon region. The discovery from Great Britain also indicates where the gladiator immortalized on the figurine studied by Alfonso Manas could have found his way to Roman arenas.

So can the statue of an "Amazon" fighting in the arena really be associated with the mythical Amazons? By collecting analyzes of several seemingly independent finds, we see that it is becoming more and more likely.

However, as we will see, this artifact seems to be associated with a mystery of even greater importance, because it goes beyond the mythical understanding of the Amazons.

What do the old messages say about the Amazons? Ancient and medieval chroniclers attributed many unpleasant features to warlike women. They were said to despise men and meet with them only to get them pregnant. Girls born of such fleeting relationships were detained by menacing women. Boys were abandoned in the wasteland to die. In addition, the Amazons reportedly cut off one of the breasts to shoot the bow more accurately. According to Greek historians, they had their own state ruled by the queen, and they got rich by plundering the surrounding lands. However, some sources emphasized the female characteristics of the Amazons, such as their beauty, their willingness to romance men, and their willingness to sacrifice themselves for the sake of their lovers.

This image of the warriors was conveyed by some Greek myths, for example, the story of the ninth of the works of Heracles, which consisted of winning the golden belt of Hippolyte, queen of the Amazons. The hero's expedition had an adventurous course, and since the queen fell in love with the hero, he easily won the belt.

However, everything ended, as it is with the Greeks, in a tragedy.

As a result of the schemes of gods and people, Hippolyte died, and Heracles fought a senseless bloody battle with her warriors. Later, during the Trojan War, the Amazons were to come to the aid of Ilion. They stood against the Greek league because their queen Penthesila was in love with the heroic Hector, son of King Priam.

Tales of the Amazons and their land circulated not only in antiquity. Female warriors also aroused interest in the Middle Ages. Adam of Bremen, a German chronicler from the 11th century, claimed that the land of the Amazons lies on the Baltic Gulf of Bothnia, and the women who live there despise men so much that when they give birth to a male child, this contempt makes the boy has a dog's head. Girls, however, do not lack anything, and even grow into beauty, because Amazons, though so bloodthirsty, are famous for their beauty. Other historiographers of that era either repeated ancient Greeks' accounts or claimed that the land of warlike women lies in the midst of the ice, northeast of Norway. As we can see, all these messages place an extraordinary kingdom on the outskirts of the world, where hardly anyone reaches, and therefore within the "wonderlands".

The news about the Amazons was also provided by Brunetto Latini, a 13th-century Florentine scientist, writer, and political activist. In his main work, The Treasury of Knowledge, in the chapter entitled On the Kingdom of Women, he wrote: "The kingdom of women came into existence when the king of the Scythians, along with all the men of his lands, died in an expedition against the Egyptians. When the Scythian women found out about this, they chose a queen from among themselves and issued a law that no man would be allowed to live in their lands. They also agreed that they would only raise daughters, not sons. They cut off their left breast to make it easier for them to carry a shield and a weapon - that is why they are called Amazons, that is, "without a single breast". They just went to help Troy. This was done by queen Penthesila, who was said to have been in love with Hector. "

The Amazon legend has come back when the New World was discovered on the other side of the Atlantic. Already in the 1630s, shortly after the Spaniards conquered the Inca Empire, a group of conquistadors set off through the Andes to a land covered with the jungle that sometime later became known as the Amazonia.

The aim of the expedition was to find and then civilize (that is, according to contemporary standards: baptism and subordination to male power) women-warriors about whom Indian fairy tales tell. The expedition did not find the Amazons.

The purpose of the expedition was to find and then civilize (that is, according to contemporary standards: baptism and subordination to male power) women-warriors about whom old Indian tales tell. The expedition did not find the Amazons.

Apparently, however, things were different in 1541, when another branch of the Spaniards, this time under the command of the famous Francisco de Orellana, looking for gold in a tropical forest, discovered the largest of South America's rivers, the Amazon River, still called the Amazon. The participant and chronicler of this brawl, the monk Gaspar de Carvajal, described how the conquistadors stumbled upon the Amazons and fought a bloody battle with them: "On June 24, we saw a village inhabited only by fair-skinned women who did not maintain relations with men. These women wore long braids, were tall and strong, and armed with bows and arrows. They attacked our ship but were repulsed, and seven or eight of them died in the fight." This description is an exception, however, because no one but Orellana's people has met the Amazons in the tropical forests of the Amazon.

But let's go back to the ancient sources. It is noteworthy that some Greek sources dealing with this topic consistently pointed to a similar location of the Amazon lands.

According to historians Herodotus (5th century BCE) and Diodorus Siculus (1st century BCE), the Amazons set off on their plundering expeditions from Scythia, the steppe areas stretching from today's Ukraine to Central Asia. The message drawn up by Herodotus seems particularly interesting, and may not be a myth, but an account of real events with the participation of warlike women. The mounted unit of Amazons was to reach the Black Sea, wherein Asia Minor, on the Termodont River, it clashed with the Greeks. The Hellenes won and then set sail on three ships with all the Amazons they had captured alive.

How many women could they imprison on the three ships? Several dozen at most. However, they underestimated the cunning, combat prowess, and desperation of the female warriors. When the ships were out at sea, the Amazons tricked themselves free, then pounced on the men and killed them. Unfortunately, they had no idea about sailing, they could not use the rudder, nor the sail, nor the oars, so after killing the sailors they got stuck on ships like in a prison.

Nevertheless, they were lucky, because the ships, carried by waves and wind, reached Kremnoj on Lake Meotsk (today's Sea of Azov), i.e. to the lands of the Scythians. The Amazons got ashore and set off on foot inland. After a few days, they encountered a group of people grazing their horses. They took the herd from them, and then they set off towards their homeland, looting the Scythians on the way, so as not to return home empty-handed. The Herodotus' message, therefore, states very vaguely that the headquarters of the Amazons were somewhere in the region of the steppes - perhaps North Caucasus, from where it was possible to reach both Asia Minor and the lands of the Scythians.

Is it true? Of course, it is not known, but it is puzzling that many ancient stories link warrior women with the steppe zone on the Black Sea, Asia Minor, and Scythia in general.

Herodotus also shows the extraordinary combat effectiveness of the Amazons. This feature was also emphasized by Diodorus Siculus when describing the expedition of an unknown ruler of the

Amazons, who, having come with troops from the east, crossed the Tanais River (today's Don) and conquered all the tribes living along the shores of the Black Sea to Thrace, i.e. the entire eastern Balkans. She returned to her homeland with enormous loot and built wonderful temples for the gods. But when it was supposed to happen, it is not known. Diodorus does not provide the location of the home of the victorious women, but it is clear that - if we treat his message as an echo of true events - they came from the eastern part of the Ukrainian-Russian steppes.

Myths and legends were needed because they contained important cultural content and instructive messages, and provided inspiration. Even today, they inspire scientists to make bold hypotheses and carry out research - and this was also the case here. It turned out that the existence of the Amazons has been confirmed by many pieces of evidence that have been buried in the graves of women warriors for millennia.

Let's start with distant Cambodia. In November 2007, a team of archaeologists from the Kyoto International Center for the Study of Japanese Culture, led by Japanese archaeologist Dr. Yoshinori Yasuda, discovered the burials of armed women in a cemetery at Phum Snay in northwest Cambodia. Previous research at the site in 1999 and 2001–2003 revealed many relics from the turn of the old and new era, but there was nothing surprising among them.

Meanwhile, Yasuda's team found the remains of 35 people buried at the beginning of the 5th century CE. Some of the skeletons were well preserved, and five of them were determined to be female based on bone characteristics. This is how the Cambodian Amazons were discovered, buried with helmets, swords, and other weapons belonging to them during their lifetime.

- Discovery of swords in women's graves is very rare. The discovery of so many burials of this type suggests that in the kingdom existing in these areas in the first centuries of our era, female warriors played an important role - Yasuda said in a statement to the AFP agency. - It should be remembered, however, that women traditionally played an important role in the former hunting and farming communities of East Asia. Only in European culture are women considered weak and in need of male protection.

Without commenting on the controversial assessments of the role of women in Asian and European cultures, one should first of all pay attention to the fact that the myth about the Amazons found material confirmation.

However, someone will ask: where is the Eurasian steppe zone, and where is the Cambodian jungle? After all, one land has nothing to do with another, located thousands of kilometers away.

However, in this case, what matters most is the confirmation of the very idea of the existence of women's armed units in the distant past. Besides, the discovery from Cambodia is not unique.

A similar find was also made in a cemetery in the city of Tabriz on the northwestern edge of Iran. At the end of December 2004, the Iranian archaeologist Alireza Hojabri Nobari showed using DNA analysis that the burial discovered years ago - considered by earlier scientists to be male - in fact, contained the remains of a woman, despite the fact that it was fully armed. So there were also warriors in ancient Iran, perhaps even leading a lifestyle as active and war-filled as the Amazons.

The Iranian scientist suspects that further studies of the remains from other graves once discovered in the local cemetery may lead to other similar findings. The archaeologist also assumes that women warriors were also buried in ancient tombs, which were recently discovered on the southern shores of the Caspian Sea.

However, the most important discovery regarding the Amazons was made in the lands where the Scythians lived long ago, i.e. in the steppe zone - where, according to Greek historians, the kingdom of the Amazons was supposed to be. In April 2001, in northwest Kazakhstan, an international team of archaeologists discovered the graves of women that explorers identify with the truest Amazons. An expedition of archaeologists from Russia, the United States, and Kazakhstan, supported by the British Royal Geographical Society, had been excavating in Kazakhstan for some time, but sensational finds were discovered only when the team reached a remote area in the upper reaches of the Tobol River.

An ancient burial ground was unearthed there, where apart from the burials of men, there were as many as seven graves of women buried with military equipment: armor, weapons, and horse harness. The bronze swords, daggers, and spear blades showed clear signs of use in line with their intended purpose in combat, so they were not only ritual equipment for graves.

Besides, although the weapons were of normal size, the handles of the slashing weapons were smaller - probably to adapt them to a woman's hand.

The assumption that the women buried in this cemetery were warriors is additionally emphasized by the tattoos covering the remains of the skin preserved on the remains. Characteristic tattoos, because according to archaeologists they have a decidedly military character. They depicted the heads of predators such as wolves and bears, as well as skulls with crossbones. The discoverers of the tombs have no doubt that these are the burials of the Amazons known from ancient Greek accounts.

"When we find a weapon in a grave with a male skeleton, we are sure that this man was involved in the war craft during his lifetime. We should reach a similar conclusion when weapons accompany women's remains," stated in an interview with the Times of Central Asia one of the members of the expedition, the American archaeologist Dr. Philip Cowell.

The Tobol River, on which the tombs of the Amazons were found, lies in a vast region called the Gate of the Peoples, because it is there, between the northern shores of the Caspian Sea and the southern Urals that for millennia successive waves of people from Central Asia have migrated to Europe.

At the turn of the 1st and 2nd millennium BCE, these areas were occupied by the Scythian tribes, some of which set off from there around the 8th century BCE west and some remained in Asiatic Scythia, as ancient writers called the eastern part of the Scythian ecumen. And it is with this culture that the graves of the Amazons are associated.

It was long believed that ancient historians indiscriminately mixed real history with mythical "history", so one should not believe their accounts that are historically or archaeologically unverifiable, such as stories about the Amazons. It is true that since the 1950s, in the areas where the Scythian and then Sarmatian culture flourished three millennia ago, single female graves with military equipment were discovered, but they were treated as isolated cases of female warriors or it was believed that the presence of weapons in these burials it resulted from ritualistic matters. This issue was summed up with short mentions, such as the one from Alexei Smirnov's work entitled "The Scythians": "The burials of young women constitute a separate group. In their graves, apart from the usual women's utensils, weapons were found; in individual cases, it is only a few arrows, in others - full gear: a quiver, two javelins knife." That's it - no conclusions, no reflection on the research problem. At the same time, elsewhere, the author barely inadvertently stated that the number of graves of women buried with swords reaches as much as 20 percent of the total number of all burials with weapons among the Sauromats. This unusual statement was also not discussed by Smirnow.

Other Soviet and then Russian scholars were also very reluctant to take up the topic of warriors of the old Indo-European steppe peoples such as the Scythians or the Sarmatians because it destroyed the established image of nomads as typically patriarchal societies.

The discovery of the tombs of the Amazons by the Tobol River changed views on this issue. Scientists began to suggest that the Amazons were not a separate tribe, but members of groups of young female warriors who formed armed bands before marriage or the birth of their first child, which trained in fighting against neighboring peoples. These groups would then be female counterparts of boys and young men called "wolves", the existence of which in many other Indo-European peoples (Greeks, Germans, Celts, Slavs) is confirmed by both written and excavation sources and mythical stories.

In view of all these finds, the hypothesis that the "she-wolves" from the Eurasian steppes formed military communities effective in combat becomes credible. It is also possible that these groups were joined by widows of fallen warriors - their sign could be tattooed in the form of a skull with crossbones, and therefore a symbol of revenge.

When were the Amazons from the Tobol River buried? The weapons found in the tombs were made of bronze, however, the

Scythians, when in the 8th century BC they reached the Black Sea through the Gate of Peoples, they were already using iron swords. So the burials of interest to us here must be earlier and come from the "Scythian Bronze Age", which lasted until the turn of the 2nd and 1st millennium BC.

The descendants of the warriors buried there probably moved to the west with the entire Scythian tribes, and also reached Asia Minor and Iran.

In the light of recent research, it seems that they could have been recorded as Amazons in Greek myths (about Heracles, about the Trojan War), and the remains of their relatives from other Scythian factions are today found in graves in Tebris or on the southern coast of the Caspian Sea.

Monuments from the burials of the Amazons from the Tobol River have been exhibited in the Kazakh capital city of Astana at the local Historical Museum.

The Amazon's tomb was also discovered in Ukraine, as reported on September 1, 2018, by the website "Ancient Origins".

In the Mamai hill in Velika Znamjanka, in the Kamjańsko-Dniprowski region in Zaporizhia, archaeologists found the burial of a woman from the Scythian culture, equipped with weapons and typically female details. In a 2400-year-old grave, a dozen arrowheads, a well-preserved bronze mirror, and lekythos, a small jar used by ancient Greek women to store aromatic oils and perfumes, were found.

"This woman was a skilled fighter," says Dr. Jeannine Davis-Kimball, in charge of the Mamai excavations. "We do not assume that she belonged to the female divisions organizing the plundering raids. We assume that she belonged to a tribe that had to protect its herds and pastures, so she was trained from childhood to fight like a man. This might have been the norm in such communities back then."

The Ukrainian Amazon lived 2,400 years ago, in an era when the ancient Greeks wrote their myths. In which on the northern outskirts of Hellas they probably encountered armed, belligerent women more than once.

Finds such as those from the Tobol River or the Zaporizhzhya River inspire scholars, but also affect the mass imagination. They intrigue and inspire not only people curious about the past but also... contemporary Amazons. For example in Ukraine. The "warriors" operating there called themselves Asgard and lived in a center hidden from the world in the Ukrainian part of the Carpathians. Asgard has over 150 people. They are mostly young women who have been hard hit by fate.

Victims of domestic violence, sexual crimes. Their leader is Katerina Tarnowska, a 30-year-old girl in white with blonde hair tied in a ponytail.

The existence of Asgard was announced to the world by Jenna Martin, a reporter for the New York-based Planet Magazine, who in the fall of 2009 set off on the trail indicated by photographer Guillaume Herbaut. Herbaut discovered an extraordinary community of women back in 2004 when he traveled to Ukraine to document the course of the Orange Revolution. The French won Asgard's trust, which allowed him to photograph and film himself.

Living in a wild region, away from other people, they reactivate the traditions of the Scythian Amazons. Of course, these "traditions" are as much based on presumptions as on the discoveries of archaeologists and the hypotheses of anthropologists regarding the ancient "wolfhounds." Ukrainian women live in a facility, the exact location of which they try to keep secret.

They learn old Ukrainian dances and all the skills traditionally considered feminine. They produce ecological food, eat traditional dishes, and - as befits the Amazons - they prepare to fight, and their main teacher is one of the few men in the community, Volodymir Stepanowycz, a karate master from the Soviet Union.

Ukrainian Amazons wear both modern clothes as well as traditional Ukrainian folk costumes. They live according to strictly defined rules that put their ideology into practice.

According to Asgard's principles, the members learn skills useful in everyday life, focusing on natural sciences, which is to make them ideal women. Living in harmony with nature, they refer to the spiritual tradition of both the Orthodox Church and the reconstructed Scythian mythology. They focus on a combination of folk art, science, sport, and health care. At the time when they were discovered for the world, Asgard did not shy away from politics, supporting the vision of the rebirth of a great and strong Ukraine and Yulia Tymoshenko as the natural leader of the nation and a strong woman.

Asgard is not a closed community, but new candidates must meet a number of conditions to be accepted. The everyday life of the members is subordinated to the ritual of working together, studying, and actively relaxing. The community includes only young women professing celibacy.

How to properly classify Asgard? Who are they?

"In my opinion, you can see at first glance that this is a sect," Guillaume Herbaut told Planet Magazine.

The founder of the community and the author of the rules in force there say something else.

"In Ukraine, many women are militant. They are proud, courageous, steadfast, and strong in faith. The spirit of the ancient Amazons still lives in our nation," assured Katerina Tarnowska in the same article published in Planet Magazine. "To bring out the best qualities in our women, they need initiation in a place separate from the rot of the world."

Ancient Amazons, those from the Tobol River, from Iran, as well as ethnically unrelated but equally dangerous in battle Cambodian "amazons", are not the first female warriors. Because understood in this way, "amazons" are ancient, older than humanity, as evidenced by the discoveries of primatologists.

The early beginnings of the "amazons" were indicated by the research of a team of scientists led by the well-known doctor Jill Preutz from Iowa State University. According to the findings of American researchers, in the group of chimpanzees, it is not males, but mainly females that use spears to hunt game. ABC News informed about these arrangements in mid-April 2015.

The herds of chimpanzees living in the Fongola area of southeast Senegal are the only creatures other than humans who regularly manufacture weapons and use them to hunt for meat. This feature caught the attention of scholars because they see it as an analogy to the behavior of ancient hominids, the ancestors of our species. According to Dr. Preutz, females usually carry their young on their backs, so they make things easier by hunting with tools.

Chimpanzees make spears from branches by removing side twigs and leaves from them, then sharpen their weapons by biting the tip with their teeth. This tool is about three-quarters of a meter long and is used to hunt small dwarf galago monkeys (Galago senegalensis), which are the staple food of chimpanzees that provide high-calorie meat and often enrich the vegetarian diet. Male Fongola chimpanzees also use such spears, but they do so much less frequently than female chimpanzees. Preutz assumes that for adult males to hunt galago, strength in arms, teeth, and aggressive behavior is sufficient. In turn, physically weaker females, burdened with young ones, make up for these deficiencies by using weapons.

"Among primates, it is females that most often find tools and use them," said the American anthropologist. "For this reason, I believe that the female invented the spears. This conclusion applies to both modern Fongola chimpanzees and the early hominids that lived millions of years ago."

So chimpanzees would be the first "amazons"?

They don't look like the slender gladiators from the Hamburg museum, but perhaps Dr. Preutz is right in saying that female humanoid species invented weapons and war.

Crystal Skull of Destiny

The story of the Skull of Destiny - as portrayed by Anna Mitchell-Hedges - could be the plot of an adventurous film. This is probably why, after many years, she inspired the creators of the popular film series about Indiana Jones to create the last episode of the series.

It all began in the ruins of Lubaantum, a Mayan city, in 1924, on Miss Mitchell-Hedges's 17th birthday.

According to Anna, her father, Frederick "Mike" Mitchell-Hedges, conducted excavations as part of the archaeological mission of the British Museum in Belize, Central America, which is rich in pre-Columbian finds. The girl accompanied him for some time. On her daughter's birthday, Fryderyk allowed her to personally work on the excavations in Lubaantun.

An excited teenager found a sculpture of a life-size human skull in a breach under a thin layer of debris made of rock crystal. As her father later claimed, the Skull of Destiny was to be crafted thousands of years earlier by the ancient Maya. This version of events was upheld by Anna Mitchell-Hedges until the end of her life.

She was steadfast, despite the fact that many researchers questioned not only the authenticity of the find but literally every information that Anna gave about it. For example, the fact that her father was not an archaeologist at all, but a journalist specializing in "sensational news," and it was not him, but Dr. Thomas Gann, who was excavating at Lubaantun. Besides, if Miss Mitchell-Hedges ever came to Belize, it would not have been until 1925. Even the date when the Skull of Destiny was discovered is not certain.

Some researchers and investigative journalists have argued that this artifact was actually discovered in Belize but in 1931 or 1939. Others, that the crystal sculpture Frederick Mitchell-Hedges had bought at an auction organized by Sotheby's auction house in London in 1943 as a curiosity of unknown origin.

Anna herself made public the controversial find in the late 1960s, sparking a wave of interest in the alleged Mayan crystal skulls - as it turned out over time that there were more such skulls. These artifacts, starting with the Skull of Destiny, capture the mass imagination. But does the legend surrounding the crystal skulls contain information that could confirm their ancient origin?

First, it is not even known how many crystal relics are attributed to the Maya, because since the presentation of the Skull of Destiny to the world, sculptures like it have started to multiply. Some say there are 200 of them, others - a dozen or so. They are made on a 1: 1 scale and weigh 3-4 kilograms. According to esoteric legends unfolding around them, these artifacts have existed for millennia, serving as "sacred storehouses of knowledge". Once they were to serve as energy amplifiers, telepathic devices, or gates to other spheres of consciousness during rituals. As esotericists claim, even today people predestined to do so can benefit from these properties of crystal skulls.

What does science say? The most famous of the crystal relics, the Skull of Destiny, has been examined by scientists several times. First, in the late 1970s, experts from Hewlett-Packard concern and confirmed that it was made of rock crystal, so it was not a quartz glass casting.

However, the tools to process rock crystal so precisely and without damaging the raw material were not invented until the end of the 19th century. So who carved this artifact, probably not the Mayans? Although they created a wonderful civilization, they could not have the tools to process a crystal, one of the forms of quartz. Modern engineers say that sculpting transparent quartz is a difficult, time-consuming task that requires great precision. So maybe Mayan craftsmen grind skulls by hand, not using tools, but sand, with which they rubbed the crystal, giving it the right form? Theoretically, it would be possible, but experts believe that creating one skull using this method would require the continuous work of many generations of craftsmen, as it would take 200 to 300 years. This finding fired the imagination of enthusiasts of the hypothesis that the Maya had in fact created a civilization that was technologically superior to ours, except that traces of their devices did not survive. Proponents of such a vision of the Mayan past had to be content with guesswork because the re-analysis of the Skull of Destiny - conducted by scientists in the late 1990s and then in 2005–2007 - showed that it was not the Maya who made the artifact. Computed tomography revealed that the quartz has traces of treatment with high-quality tool steel tools. Scientists had no doubt that the famous skull was created between 1890 when such tools were introduced, and 1969, when Anna Mitchell-Hedges showed the artifact to the world.

The legend has fallen. The Skull of Destiny was not made by the Maya, it was not created thousands of years ago in an unknown place and in mysterious circumstances. So who and for what purpose made this artifact, involving a lot of energy and resources?

The mystery of the crystal skull and its imitations that are difficult to count still await clarification.

About the author

Tadeusz Oszubski, born in 1958 in Bydgoszcz. Prose writer, poet, publicist, journalist, cultural animator. He belongs to the Polish Writers' Union.

He made his debut in May 1980 with a poem published in the literary biweekly Kamena (No. 10/1980). His poetry has been published in many magazines. In 1984, Oszubski's collection of poems "We left our wings in the cloakroom" (K.P.T.K., Bydgoszcz 1984). The writer is known primarily as a prose writer and author of popularizing books.

He writing works from the border of crime fiction, thriller, noir literature, horror, and fantasy, as well as contemporary, literary adaptations of traditional legends from various regions of Poland. His novels are: "Pack" (Almanach of Horror and Fantasy "Voyager" No. 3, Publishing House Przedświt, Warsaw 1992), nominated for the Polish Fandom Award. Janusz A. Zajdel in 1992; "Messiah" (Kornel Ceglarski Publishing Agency, Bydgoszcz 1993); "Stronghold of hate" (S.R. Publishing House, Warsaw 1997); "Love" (Erica Publishing Institute, Warsaw 2015); "Bydgostia" (published by Project Legendaria, Bydgoszcz 2017); "Young" (Łuczniczka Publishing House, Bydgoszcz 2017). The collections of Oszubski's stories include "Predator" (published by Świat Książki, Warsaw 1999), awarded in the 2nd National Competition for Polish Contemporary Novels; "Goplana's tears". Stories based on the motifs of Polish legends about aquarists, water ladies, and mermaids (Biblioteka "Tematu", Bydgoszcz 2016); "City" (Łuczniczka Publishing House, Bydgoszcz 2018); "ZakoHanna" (Łuczniczka Publishing House, Bydgoszcz 2019); "Legend of the Water Maidens" (Łuczniczka Publishing House, Bydgoszcz 2019), "Interregnum. Stories "(Łuczniczka Publishing House, Bydgoszcz 2020).

Tadeusz Oszubski is also the author of popularizing books: "Mysterious Creatures" (KOS Publishing House, Katowice 2004, translation into Czech "Zahady z rise zvirat", Alpress, Prague 2006); "Mysteries of the world" (published by KOS, Katowice 2005) and "Mysterious artifacts" (Erica Publishing Institute, Warsaw 2016), recognized in the annual plebiscite of the Granice.pl portal for the "Best book for Christmas 2016" in the "For history lovers" category. His reporter's book "Unexplained Phenomena in Poland" (Videograf II, Katowice 2003), written together with Wojciech M. Chudziński, was nominated for the Grand Press 2003 nationwide journalist award.

Over 50 of the author's stories have been published in anthologies and magazines. The most important prints in the anthologies: "Homunculus Mirabilis" in "Maladie" (an anthology of Polish fantasy prose for the Czech market, translation into Czech, published by Leonardo, Ostrava 1996); "Interregnum" in "Black Mass" (Rebis, Poznań 1992); "Machine" in "Unique offer. The best film stories "(Machina, Warsaw 2000)," Romek "in" City3. An anthology of Polish horror stories" (Forma, Szczecin 2016)," Ptakot "in" City4. An anthology of Polish horror stories "(Forma publisher, Szczecin 2018),

Some of the stories published in magazines are: "Hobby of an elderly gentleman" (weekly "Kujawy i Pomorze" September 1991); "Children of the Rifle" (Przegląd Artystyczno-Literacki 1998 No. 12); Machine (Artistic and Literary Review 1998 No. 12); "The fearless witch hunter" (Przegląd Artystyczno-Literacki 1999 No. 5; Fahrenheit No. XXXII 2003; Action Naked Thursday No. 10/2005); "Substation" (Phrase 1998 No. 19/20); "The Girl Behind the Mirror" (Phrase 2001 No. 3); "With Coppelia in my heart" (Phrase 2004 No. 1); "Knight of the Christmas Eve" (Nowa Fantastyka 1994 No. 3); "Pillars of Heaven" (Nowa Fantastyka 1995 No. 7); "Homunculus Mirabilis" (Voyager 1993 No. 5); Lord of the Dungeons (Voyager 1992 No. 2); "Underneath" (Fenix 1994 No. 2); "Więź" (Fenix 1995 No. 8); "Moisture" (Fenix 1996 No. 11); "Cool Time" (Fenix 1997 No. 3); "The House of Orchids" (Fenix 1997 No. 9); "Betrayers of the Nations of the World" ("Brainstorm" Summer 1997); "Goodbye, doll" (Fahrenheit, issue XLIII - December 2004; Naked Thursday Action No. 5/2005); "That wonderful Saint-Germain!" (Naked Thursday Action No. 9/2005); "Gotyk" (Science Fiction 6/2004); "In the Name of a Cat" (Science Fiction 10/2004; Action Naked Thursday No. 8/2005); "Behind the Wall" (Naked Thursday Action No. 7/2005); "Second cage" (The Quarterly. Irregular Pismo Kulturalne No. 3/2011); "In Helena's Head" (Naked Thursday Action No. 14/2005); "Living Steel" (Naked Thursday Action No. 6/2005); "Headhunters" (Naked Thursday Action No. 12/2005); "Captive" (Naked Thursday Action No. 11/2005); "Second cage" (Naked Thursday Action No. 13/2005; "bazaebokow", December 2014); "Golem" (Naked Thursday Action No. 15/2005); "The Pact" (Fahrenheit No. XXVIII, June 2003); "Blood" (Naked Thursday Action No. 17/2006); Baghdad's Most Beautiful Woman (The Fourth Dimension No. 6/2000); "Paprotek" (The Fourth Dimension No. 2/2000); "Avatar" (Fahrenheit No. XXX August 2003;

"bazaebokow", December 2014); "Red Wall" (SFera No. 1/2003); "Serial loneliness" ("bazaebokow", December 2014), "Under the bridge" (OkoLica Fachu, June 2016; Nagi Thursday Action No. 16/2006), "Whirlwind of love" (Theme, Autumn-Winter 2016); "Romek" (Temat, Autumn-Winter 2016); "The Bones of the Fathers" (audiobook "Arsonists of the Sky", March 2018).

The writer has been awarded in many literary competitions. Among others: For a Wooden Owl (Gostyń 1980), 2nd Stanisław Grochowiak Poetry Competition (Radom 1981), Competition André Norton for a novel and short story (Poznań 1991), Literary Competition Stefan Themerson (Płock 1993), Competition for a story and audio By the Odra and Baltic Sea (Szczecin 1994), Polish Contemporary Novel Competition (Warsaw 1998), Film Short Story Competition (Warsaw 1999). In 2019, the Mayor of the City of Bydgoszcz awarded Tadeusz Oszubski with an annual Artistic Scholarship for an outstanding creator of culture.

Since 1991, as a reporter, publicist, editor, Oszubski was an employee and collaborator of many regional and national press titles, including "Illustrated Polish Courier", "Express Bydgoski", weekly "Kujawy and Pomorze", monthly magazines "Fourth Dimension", "Unknown World".

Tadeusz Oszubski is the co-founder of the Foundation Our Tradition - Our Future (the seat of the Foundation is in Bydgoszcz), its creative director, and the chairman of the Foundation's Supervisory Board. In addition, he is the author of the Legendaria Project promoting Polish legends and a co-organizer of other cultural and educational activities of the Foundation, including those carried out at the Legend House of Bydgoszcz, the cultural center of the Foundation Our Tradition - Our Future.

Table of Contents